Neuroscience of Pornography Addiction

Kendir Ramiz

Neuroscience of Pornography Addiction

Introduction

We live in an age of unprecedented access, a time where the most obscure knowledge and the most intense sensory experiences are but a few taps away. The digital realm has woven itself into the very fabric of our existence, transforming how we communicate, learn, and yes, how we experience pleasure. Among the countless offerings of this digital landscape, pornography stands out as a particularly potent force—a phenomenon that is simultaneously pervasive and yet, shrouded in a strange sort of silence. It's a silent giant, living within our devices, in our pockets and on our desks. It's a whisper in the ear of our most basic desires, a flickering screen that can captivate our attention, and alter our brain's architecture in subtle and not-so-subtle ways.

Kendir Ramiz

The sheer accessibility of pornography today is nothing short of revolutionary. Gone are the days of clandestine searches in dingy corner stores or hidden magazines under the mattress. Now, with a device that fits in the palm of your hand, an individual can, with just a few clicks or taps, access an unending stream of explicit content from the privacy of their home, on the bus, or even during a break at work. This ubiquity, this constant presence, is not merely a matter of convenience; it's a fundamental shift in how we relate to our own sexuality, and how we perceive the bodies and desires of others. Pew Research Center, in its various studies on internet usage, has consistently documented the exponential growth of online activity, and while specific statistics on pornography viewing are often difficult to obtain due to the private nature of the activity, the general trends and associated data are clear: the consumption of digital content, including sexually explicit material, is a dominant form of engagement in the 21st century. This isn't merely a technological evolution; it is a cultural and neurobiological transformation occurring right before our very eyes.

Neuroscience of Pornography Addiction

But this book isn't about judgment; it's not about finger-wagging or puritanical condemnation. We've seen far too much of that, and it has never yielded particularly useful results. Instead, we're going to shift our gaze away from the stale arguments of morality and delve into something much more profound—the neurological and psychological processes that are triggered by the experience of pornography. We'll be setting aside the heated rhetoric and engaging in the quiet work of observation and analysis. We're not here to tell you what you "should" or "shouldn't" be doing; we are here to unpack the science of what is happening inside your brain when you engage with these experiences. This isn't about labeling behaviors as inherently "good" or "bad" but rather about understanding the complex dance of chemicals, circuitry, and conditioning that shapes our human behavior. We will explore how the brain responds, adapts, and sometimes, succumbs to the potent influence of such ubiquitous stimuli.

Kendir Ramiz

The perspective we'll adopt throughout this exploration is decidedly neurochemical. We'll be looking at the brain not as some monolithic entity, but as a complex ecosystem of neurotransmitters, hormones, and neural pathways—all interacting in a continuous dance of give and take. We will acknowledge that pornography addiction isn't a moral failing or a matter of weak willpower; it is, at its core, a neurobiological phenomenon. What does that mean? It means that the brain's architecture and functioning are physically altered by repeated exposure to certain stimuli—in this case, explicit content designed to trigger the most potent reward pathways. It's about understanding that there are actual, tangible, chemical changes occurring within the brain when a person develops a compulsive relationship with pornography. These alterations are not just theoretical constructs; they can be observed and measured using tools like functional MRI and other imaging techniques. These tools allow us to glimpse the hidden mechanisms of our own desires, and in that unveiling, we might discover the path to more helpful solutions, to treatments and preventive measures that are built on a firmer foundation of understanding.

Central to this discussion is the brain's reward system — a remarkable piece of biological engineering that has, throughout our evolutionary history, ensured our survival by motivating us to engage in behaviors necessary for our existence. The mesolimbic pathway, particularly, has been a main focus in understanding addiction in general and has shown the role dopamine plays in the reward system. This is the circuitry that floods us with a sense of pleasure when we eat a satisfying meal, experience intimacy, or achieve a long-sought goal. But, in the case of pornography, this system is often hijacked. The sheer intensity and novelty of the content often bypasses normal reward mechanisms, producing a far more powerful dopamine surge than what would be triggered by real-life experiences. The curated nature of pornography—the perfect bodies, the heightened emotional responses, and the continuous stream of novelty—creates a kind of supernormal stimulus, one that has the potential to overwhelm and ultimately alter the brain's natural reward circuitry. This is the root of why this behavior can move from a casual experience to a compulsive behavior, why that initial pleasure-seeking can turn into an insistent and at times maddening need for more. In real life, we don't experience this level of curated and intense engagement all the time, which means that pornography is, in a very real sense, a kind of unnatural stimulant for our neural reward system.

And this brings us to the core of why this deeper understanding is so important. When we are armed with scientific literacy about the processes underlying addiction—not just pornography addiction, but all forms of compulsive behavior—we can move beyond the simplistic, often unproductive, approaches. We can begin to develop more effective treatment and prevention strategies. When we comprehend how the brain changes under the influence of these intense stimuli, we can design interventions that target the underlying neural mechanisms, rather than relying solely on willpower or moral persuasion. The ability to discuss these issues openly, honestly, and without judgment creates an environment where people feel empowered to seek the help that they need. This scientific literacy also fosters a sense of agency; it shows that individuals are not simply passive recipients of their own desires, but can actively participate in shaping their own neurochemical landscape. It's not about absolving individuals of responsibility, but rather, about providing them with the knowledge they need to make more informed choices, and to reclaim their own neural autonomy. This understanding is not just about treating the addicted; it is also about creating a society in which everyone is empowered to live and flourish with a deeper understanding of their own neurochemical landscape.

This book is an invitation to embark on a journey of discovery—a journey into the inner workings of the human brain, and a quest to understand the complex relationship between desire, compulsion, and the neurochemical world that shapes us all. We are beings of both reason and instinct; and this is about bridging that gap so that we can both appreciate the power of our instincts and still make good choices along the way. It's about transforming our relationship with our desires, and understanding how we are built, wired and operating so that we can engage with life from a position of wisdom and empowered action. And in so doing, we might just uncover new paths towards wholeness, healing, and a richer more empowered life. This isn't just about science; it's about living a more conscious life, and as a wise individual once said, "The unexamined life is not worth living." So let's examine, let's explore, and let's move forward with a sense of curiosity and hope.

Neurotransmitters and Hormones in Action

Let's delve now into the very genesis of desire, the first flickering of arousal that sets the stage for the complex interplay of events we will discuss in the chapters to come. It's like the moment the curtain rises on a grand theatrical production—a flurry of activity backstage, invisible to the audience, yet essential to the unfolding drama. This initial spark, this surge of sensation, is not just a simple, singular event; it's a symphony of chemical reactions orchestrated within the labyrinth of the human brain. It's a cocktail of neurotransmitters and hormones, each playing its unique and indispensable role in creating that intoxicating experience of sexual arousal.

Kendir Ramiz

We often think of arousal as a purely physical sensation, a tingling in the body, a quickening of the pulse, a flush on the skin. And indeed, these physical manifestations are certainly present, but they are simply the outward expressions of a far more intricate, internal process. Deep within the brain, a coordinated release of powerful chemical messengers is occurring. This is not a static event, it's a dynamic, ever shifting cascade of chemical signals, each triggering the next, creating that feeling of intensity and urgent desire. Think of it as an orchestra tuning up; each instrument begins to play its notes, each adding layers and complexity to the overall sound. So what exactly are the instruments of this orchestra? What are the key players in this initial dance of desire?

The first, and perhaps most well-known member of this chemical orchestra, is dopamine. This neurotransmitter is often referred to as the "pleasure chemical", and while that's a simplification, it's not without merit. Dopamine is a key player in the brain's reward system, the circuitry that motivates us to seek out and engage with

Neuroscience of Pornography Addiction

experiences that are crucial to our survival and well-being. It's not just about feeling pleasure but it's also about motivation and the anticipation of pleasure. Dopamine doesn't merely register pleasure after an event, but, it is released as we seek out or even just anticipate pleasurable experiences. In the context of sexual arousal, dopamine surges when we are exposed to a sexually stimulating image, a seductive sound, or a tantalizing fantasy. This initial surge is like the lighting of a fuse, setting off a chain reaction that amplifies the overall sense of desire. The work of Professor James Pfaus, detailed in his 2009 paper "Pathways of Sexual Desire," illuminates the pathways through which these dopamine-mediated responses occur, specifically highlighting the role of the mesolimbic pathway, a key player in the brain's reward circuit. This pathway, like a superhighway of desire, carries messages between the brain's pleasure centers, reinforcing the behaviors that trigger dopamine release and thereby strengthening the drive to seek it out. The insights of Wolfram Schultz and his work on "Dopamine Reward Prediction Error Coding" further highlight how dopamine isn't just a reward signal, but also a signal that encodes anticipation and surprise, further solidifying the power of this neurotransmitter in driving our behavior. It's as if our brains are constantly recalibrating the potential for pleasure, making the pursuit of that experience all the more alluring.

Dopamine does not act alone. Another crucial neurotransmitter involved in this initial spark is glutamate. While dopamine often gets the limelight, glutamate is its essential partner in this complex dance. Glutamate is the most abundant neurotransmitter in the brain, and it plays a crucial role in learning and memory. It's the essential neurotransmitter in many types of neural plasticity and is necessary for long term potentiation and long term depression, the mechanisms the brain uses to either strengthen or weaken connections. In the context of sexual arousal, glutamate acts as a kind of amplifier, strengthening the associations between specific stimuli and the subsequent feelings of pleasure. In simpler terms, it helps the brain learn what things are sexually arousing to an individual. When combined with a dopamine surge

Neuroscience of Pornography Addiction

this is a recipe for a powerful habit. Glutamate is essential in the process of forming new neural pathways which can then be reactivated with that initial spark of arousal, making the connection between specific stimuli, for example a certain type of image and arousal, that much more powerful. This process is explained in the work of Lüscher and Malenka, who discuss the crucial role of NMDA receptors (one type of receptor for Glutamate) in their study, "NMDA receptor-dependent long-term potentiation and long-term depression (LTP/LTD)," they detail how these processes shape the way the brain learns and remembers, underscoring how these neurological mechanisms contribute to the development of habits around sexual behavior.

Beyond these neurotransmitters, a complex array of hormones is also released as that initial spark of desire ignites. Among these hormones, testosterone stands out as a particularly important player. Often associated primarily with male sexuality, testosterone plays a crucial role in both men's and women's sexual desire, and overall well being. It is not simply a "sex hormone" but a regulator of a wide range of bodily and cognitive functions, including muscle mass, bone density, and mood. But in the context of sexual arousal, testosterone's primary role is to amplify the pleasure response. As detailed in the "Andro Accelerator Hypothesis," adding testosterone to the mix of neurotransmitters and hormones intensifies the overall sense of desire, making the experience more rewarding and more compelling. This does not operate in a vacuum, of course, and is directly related to the influence of the other neurotransmitters like dopamine and glutamate. This interplay explains why those with higher testosterone levels (both men and women) often experience a more intense motivation for sexual activity, but it's also important to recognize that the levels of testosterone in the body are very fluid, changing based on a variety of factors including time of day, diet, physical activity, and mental and emotional states. So while testosterone plays an important role, it does not solely dictate an individual's desire or behavior. The work of Sari van Anders and her examination of the link between testosterone and sexual interest in women further highlights the complex ways in which hormones impact sexual behavior, challenging the often simplistic notion of testosterone as solely a "male" hormone.

There are other key hormones involved. Oxytocin, often referred to as the "love hormone," is also released during arousal and has a role in pair bonding and social connection. Endorphins, the body's natural painkillers, contribute to the overall sense of well-being and euphoria, adding another layer of complexity to the experience of sexual desire. And nitric oxide, though not a hormone in the strictest sense, is a key molecule that helps with vasodilation which leads to the physical manifestations of arousal. These various molecules all contribute in their own way to the overall picture of sexual arousal. It's not just about what happens in the brain but how those processes are then expressed in the body. The feeling of arousal, that familiar rush of sensation, is a reflection of this complex choreography occurring throughout the body.

Now, you might be asking, what about the stress response? How does that fit in? The HPA axis, short for the Hypothalamic-Pituitary-Adrenal axis, also springs into action as desire begins to take hold. This is the body's stress-response system, designed to release cortisol and other stress hormones that prepare the body for action. When you feel a craving, or a need for sexual stimulation, the HPA axis kicks into gear, providing the necessary energy and motivation to pursue your desire. This is why it can be so difficult to resist that pull, it is as though the entire body is being primed to engage in that behavior, and the initial desire or thought is reinforced by this cascade of stress hormones. The work of Herman and Cullinan, "Neurocircuitry of stress: central control of the hypothalamo-pituitary-adrenocortical axis," details the mechanisms by which the HPA axis operates, further highlighting the interplay between our stress response and our desires. This response is not just limited to the pursuit of sexual gratification; it's a fundamental aspect of how our bodies are wired to seek out resources, rewards, and pleasurable experiences. It explains why cravings are so difficult to dismiss, and why, at times, our bodies seem to have a will of their own.

So, what does all of this mean? It means that the initial spark of arousal is far from a simple, isolated event. It's a rich, layered experience, driven by a precise symphony of neurotransmitters and hormones, all working in concert to create the potent sensation of sexual desire. It's an orchestra of molecules and impulses, each playing its unique part in the overall experience. It's important to recognize that this process is highly complex and not yet fully understood, so, we are always learning more. By digging into the details of this process we gain a richer understanding of our own experiences, and the forces that drive our actions. We can see that what might appear on the surface to be simply "wanting," is actually a complex dance of biology, psychology, and habit. As we journey further into this book we will be unpacking the complex ways in which this initial spark interacts with the brain's reward system to create the conditions for habitual and, at times, compulsive behaviors. We will see how these initial chemical reactions contribute to the powerful pull of pornography, and in turn, shape the very architecture of our brains.

This, then, is just the beginning—the tuning of the instruments, the first notes of the symphony. As we move forward, we'll continue to explore the various layers of this complex process, understanding how these initial reactions lead to the formation of habits, the development of tolerance, and ultimately, the potential for both addiction and recovery. But for now, let's simply acknowledge the beauty and complexity of this initial spark, this rich and dynamic cocktail of chemicals that sets the stage for all that is to follow.

Neuroscience of Pornography Addiction

From Pleasure to Habit

Having explored the initial spark of desire—the complex interplay of neurotransmitters and hormones that ignite our senses—it is now time to delve deeper into the architecture of habit. Because, as all of us know, human beings are not simply creatures of fleeting impulses, we are also, profoundly, creatures of habit. Our daily routines, our preferences, our deeply ingrained behaviors, are all testament to the powerful and pervasive influence of the brain's reward circuitry. It is this circuitry that takes an initial experience, one that may have started as a novel and exciting event, and transforms it into something that feels as if it's built into our very being. The journey from a first encounter with pornography to a deeply ingrained habit is, at its core, a journey through this incredibly intricate and often underestimated system. This chapter will be looking at how the brain transforms a sensation into a compulsion.

Kendir Ramiz

At the very heart of this process is the mesolimbic dopamine pathway. This isn't just a mere anatomical structure, but a vibrant network of neurons, a superhighway connecting the brain's most primitive regions to its more sophisticated areas. It's a system that has been honed over millennia through evolution to guide our behavior. This pathway, beginning in the ventral tegmental area (VTA) and extending to the nucleus accumbens (NAc), is a key player in experiencing pleasure, motivation, and reward. When we encounter something pleasurable, that initial spark we discussed in the previous chapter initiates a cascade of events, triggering the release of dopamine along this pathway. The VTA serves as the initial sending station, firing signals that cause dopamine neurons in the NAc to release their chemical messengers. This surge of dopamine, is not just a mere sensation; it's a powerful signal that reinforces the behavior that preceded it. It's the brain's way of saying, "Pay attention, this is important. Do this again." The work of Roy Wise and his 2002 paper "Brain dopamine and the rewarding value of

Neuroscience of Pornography Addiction

addictive drugs" provides an important framework for understanding this process. Wise details how the mesolimbic pathway is a common mechanism for many behaviors, addictive and otherwise. These studies demonstrate that the same neural circuits are activated when someone experiences a pleasurable activity, takes an addictive substance, or even engages in a behavior like gambling. This highlights just how powerful these systems are in shaping human behavior. Wise's research goes beyond merely acknowledging dopamine as a "pleasure chemical," emphasizing its role in assigning motivational salience, the brain's way of flagging specific events as important and worth repeating. This makes it a driving force in the formation of habits.

This brings us to the formation of habits, one of the most powerful mechanisms by which we, as humans, structure our daily lives. Habits are not simply mindless routines, but rather, deeply ingrained patterns of behavior that are driven by the underlying neurochemical processes. The brain, ever the efficient machine, favors established patterns; it seeks to reduce cognitive load by making repetitive actions more automatic. When we engage in an action repeatedly, and especially when those actions are associated with a strong dopamine surge, the brain begins to strengthen the neural connections associated with that behavior. The basal ganglia, a region of the brain involved in motor control and habit formation, plays a key role in this process. With repetition, a behavior shifts from being a deliberate choice to an automatic response, a sort of autopilot mode that requires less conscious effort, freeing up cognitive resources for other tasks. As Ann Graybiel brilliantly explores in her 2008 paper "Habits, rituals, and the evaluative brain," habits are not simply a collection of mechanical actions, but are rather, deeply ingrained patterns that shape our expectations, our routines, and ultimately our identities. Graybiel's work delves into the complexity of habit formation and explains how initially effortful or rewarding activities transition to automated routines that are largely carried out with little or no conscious thought. These patterns allow us to perform complex tasks with ease but, they

Neuroscience of Pornography Addiction

also make it very difficult to change deeply ingrained behaviors, whether those habits serve us well, or ultimately lead us astray. It is this very mechanism that can make it challenging to break free from a habit once it has been established. Once a behavior is encoded in the basal ganglia it's as if it's a kind of autopilot that's constantly working beneath the surface. This is why people who are actively trying to change a habit, often still find themselves acting out the old pattern.

Now, consider the implications of this reward pathway in relation to pornography. When we are exposed to an image or video designed to elicit a strong sexual response, the mesolimbic pathway activates, flooding the brain with dopamine. With repetition, the brain begins to associate specific images, sounds, and scenarios with this dopamine surge, strengthening the neural pathways and making the desire to engage with that material more insistent. This is not merely a mental phenomenon, it's a physical rewiring of the brain. But there is an added factor here that amplifies this effect: the reality that much of the pornography available today is designed to be a "supernormal stimulus." What does that mean? A supernormal stimulus is an exaggerated version of something that triggers our natural reward pathways, leading to a much more intense reaction than what we would experience in the natural world. In our evolutionary history, we have been rewarded for certain behaviors, like eating a sweet fruit or experiencing intimacy with another person. These behaviors trigger the release of dopamine, reinforcing the drive to seek these experiences. However, in our current world, we have created these artificial versions of these same experiences. Pornography is not real intimacy, and it's not real experience, but the brain processes it as though it is. It is curated and edited, making it seem far more perfect, pleasurable and readily accessible than is possible in the real world. The faces are often more attractive, the bodies more sculpted, and the sexual encounters often far more theatrical than one would encounter in a typical interaction. This makes the brain believe this to be an even more intensely rewarding experience. It's a shortcut, a hyper-realistic depiction of desire that often bypasses the natural controls, and has

Neuroscience of Pornography Addiction

the potential to overwhelm the reward system with an excessive amount of dopamine. This leaves the individual craving an impossible and ultimately unreal experience. This creates the condition for a tolerance to grow to these hyper-realistic stimuli, making normal stimuli feel unsatisfying.

This leads to the idea that the brain is always looking for the "path of least resistance." It's not inherently lazy, but rather, designed to conserve energy and streamline processes. It will often favor the path that offers the most immediate reward with the least amount of effort. Pornography offers instant access to high levels of stimulation, requiring minimal effort or social risk. In a world where real intimacy requires vulnerability, emotional investment, and often complex social navigation, pornography often presents itself as the easy alternative. The brain, not designed to differentiate a real experience from a simulated one, is simply responding to the dopamine surge, reinforcing the behavior that leads to that immediate reward. This is not to say that real intimacy is more difficult or bad, but, when your brain has been exposed to intense, readily accessible versions of experience, the brain may naturally prefer the easier option because it requires less effort, and is ultimately, more predictable. This can lead people to favor a digital experience with a screen over the real experience with another person, often without even realizing it's happening. This is not a moral failing, it's a predictable outcome of our brain's design.

It's important to discuss the role of classical conditioning. This is a learning process where a neutral stimulus becomes associated with a specific response through repeated pairing. In the case of pornography, specific images, sounds, or scenarios may become associated with the surge of dopamine, creating a conditioned response. Over time, even simply seeing a website logo, hearing certain music, or reading a specific word may trigger a craving for that dopamine hit. Pavlov's dogs famously salivated at the sound of a bell after they had been repeatedly conditioned to associate it with food, and a similar process happens in our brains regarding pornography. It's a conditioned response that the body learns to associate with the dopamine hit, creating a craving that often seems to come out of nowhere. This highlights how powerful and enduring these neural associations can become, and is a clear demonstration of how these patterns become deeply ingrained and hard to change.

The journey from an initial experience of pleasure to a deeply ingrained habit is, at its core, a journey through the brain's reward circuitry. The mesolimbic dopamine pathway, the basal ganglia, the impact of supernormal stimuli, the tendency towards the path of least resistance, and the power of classical conditioning all interplay to transform a simple behavior into a powerful, at times overwhelming habit. By understanding these underlying mechanisms, we begin to see how the brain transforms an initial experience of pleasure into a deep-seated compulsion. We start to grasp why it's so difficult to break free from the pull of these habits, and we begin to see the necessity of understanding not only the initial spark, but the systems that transform that spark into a self-perpetuating cycle. We move on now to explore what happens when these systems begin to malfunction and how the brain, in its attempt to self-regulate, creates even more challenging obstacles on the road to recovery.

Excitotoxicity and the Brain's Balancing Act

Having examined the reward circuitry and the formation of habits, we must now confront a more complex and potentially damaging aspect of the neurochemical landscape: excitotoxicity. This is where the brain's attempt to create balance can, if unchecked, actually lead to self-inflicted damage. It's not that the brain doesn't try to keep itself healthy, but its very methods of self-regulation can, if driven too far, begin to cause problems. This chapter is about understanding how the brain can, in its attempt to manage overstimulation, actually contribute to the downward spiral of addiction. It's about understanding the delicate balance of neurotransmitters, and how that delicate dance can go wrong when pushed too far. The brain is not a simple machine; it's a complex ecosystem, and as such, when there is too much of one thing, or not enough of another, the whole system starts to suffer.

Kendir Ramiz

The term "excitotoxicity" itself sounds rather dramatic, and, in many ways, it is. It refers to a phenomenon where excessive stimulation by certain neurotransmitters, most notably glutamate, can lead to cellular damage and even death. Glutamate, as we discussed in the previous chapter, is essential for learning and memory, acting as a kind of excitatory accelerator in the brain. However, when glutamate is released in excessive amounts, or when it is not properly regulated by inhibitory mechanisms, it can become toxic to neurons, causing them to become over-stimulated and eventually die. This process is akin to turning up the volume on a sound system too high; initially the sound may be exhilarating and exciting, but with time that excessive stimulation, that constant bombardment, is damaging the system. This is like trying to keep a car engine running in the red zone, or trying to bake something at too high of a temperature: it will eventually burn itself out. This is what happens in the brain when excitotoxicity takes hold. The research of Dennis Choi, particularly his 1988 paper "Glutamate neurotoxicity and diseases of the nervous system," is seminal in understanding this process. Choi's work detailed how excessive glutamate signaling can trigger a cascade of events that disrupt cellular function, eventually leading to cell death through what has been described as an "excitotoxic" process. It is important to recognize that this is not merely a theoretical construct, but a very real phenomenon that has been observed in various neurological conditions, including stroke, traumatic brain injury, and neurodegenerative diseases.

To conceptualize this, let's visualize the interplay between excitatory and inhibitory neurotransmitters as opposing "hot" and "cool" states within the brain. Glutamate, along with dopamine and other excitatory neurotransmitters, represents a "hot" state, one that is characterized by increased neuronal activity and heightened responsiveness to stimuli. When we experience sexual arousal or the anticipation of pleasure, the brain becomes "hot," with increased activity in the mesolimbic pathway. This heightened state is not inherently harmful; it's an essential component of experiencing pleasure and motivation. However, if this "hot" state is sustained for too long, or if it's triggered too frequently, it can start to become destructive. Think of a car engine that's always running at full speed, it won't take long before things begin to overheat. On the other side of this equation, we have the "cool" states: where inhibitory neurotransmitters, and regulatory hormones help the system return to a state of homeostasis, reducing excitation and promoting relaxation. These calming signals prevent the brain from becoming overwhelmed. It's this very balance between "hot" and "cool" that's critical for maintaining healthy brain function. We need both states, but when one overpowers the other, problems arise.

Let's specifically discuss how this applies to pornography. As we established in previous chapters, the intense and curated nature of pornography triggers an excessive release of glutamate and dopamine, creating a "hot" state in the brain. This initial surge is pleasurable, but with repeated exposure, the brain begins to struggle to manage this level of excitation. The

Neuroscience of Pornography Addiction

brain's own innate mechanisms for maintaining balance start to falter. This isn't simply a matter of willpower or lack of self-control, the brain's very architecture is being impacted by the repeated exposure to these intense stimulations. So what does the brain do in an attempt to try to get back into balance? Well, it starts to release the "cool" signals.

Enter prolactin. Often discussed in terms of female reproduction, prolactin also plays a crucial role in male sexual physiology, and in the overall regulation of the brain. It is released after orgasm, and as such is a key player in the refractory period, the time after sex when it is difficult to get aroused. Prolactin acts as a sort of brake, helping to cool down the system and return the brain to a more balanced state. It's as though, after an intense workout, the body starts to send signals of relaxation and recovery, and the release of prolactin is part of that process. While it is not necessarily a calming neurotransmitter, it plays an important role in the recovery process, preventing the system from staying in that "hot" state indefinitely. But that's not the only player in this regulatory dance.

Another significant player in this braking system is dinorphin. This is a type of neuropeptide, a chemical messenger that's released in response to excessive dopamine release, and serves to dampen the effect of dopamine. While dopamine excites and motivates, dinorphin essentially acts as a damper, reducing that feeling of pleasure and motivation. Its action is a kind of corrective measure, a message from the brain that enough is enough. It signals a sort of dissatisfaction that starts to creep in when the individual has pushed things too far. As explored in the research of Shippenberg and Herz in their 1988 work "Motivational properties of opioids: the role of the kappa-receptor system," dinorphin interacts with the brain's opioid system, a crucial part of the reward network, and can actually reduce dopamine's effectiveness as the system attempts to regain its balance. Its purpose, in short, is to make the experience of seeking out dopamine less rewarding. This is not a pleasant experience, it is in

Neuroscience of Pornography Addiction

many ways the beginning of the addictive cycle, and the brain is now entering a phase of trying to reduce the amount of pleasure felt, and this often leads to a feeling of frustration, dissatisfaction and an urge for more. It's not that dinorphin is trying to be cruel, it's trying to restore balance to a system that has been pushed to its limits, but, unfortunately, in the long run, its actions contribute to the downward spiral of addiction.

These compensatory mechanisms, designed to protect the brain from excitotoxicity, can create other problems. When the brain is repeatedly subjected to excessive stimulation, it struggles to regulate itself, and ultimately begins to alter its own architecture. This is where we encounter the phenomenon of "hypofrontality," a condition where chronic overstimulation, and the brain's attempts to compensate, leads to a dysfunction in the prefrontal cortex. The prefrontal cortex, located in the front of the brain, is a crucial region for higher-order cognitive functions such as planning, decision-making, impulse control, and emotional regulation. It's often referred to as the "CEO" of the brain, the area that is responsible for directing our actions and making executive decisions. When the prefrontal cortex is impaired, decision-making becomes erratic, impulse control weakens, and the ability to regulate emotions diminishes. It is as though the executive office has become disorganized and unable to direct the other departments within the brain. This is not a trivial matter; it fundamentally changes the way a person behaves. As these higher-order executive functions are slowly eroded by the chronic overstimulation, the result is that the individual starts to respond more from those more impulsive lower regions of the brain, instead of the decision making higher regions. It is at this point that addiction is not only a habit but a compulsion, and the individual loses the ability to make good decisions and exercise self-control. This is not simply a lack of willpower, it's a neurological shift in the way the brain operates.

Excitotoxicity and the brain's attempt to counter it creates a complex and often self-defeating cycle. The initial surge of pleasure, driven by the over-release of

Neuroscience of Pornography Addiction

neurotransmitters, triggers a cascade of events aimed at restoring balance, but, those very attempts at balance can contribute to further dysfunction. The brain, attempting to cool itself down, sets in motion a process that ultimately impairs its ability to function effectively. This demonstrates that the process of addiction is not about a lack of self control, but that our brain's very defense mechanisms begin to work against us. The constant push and pull between excitation and inhibition begins to take its toll, leading to a diminished capacity for reason, impaired impulse control, and ultimately a deep and often isolating sense of struggle.

So where do we go from here? If the brain, in its attempt to regulate itself, begins to malfunction, is all hope lost? Absolutely not, and that is the good news. Understanding these processes, is our first step in creating a more conscious relationship with our behaviors. It's about recognizing that the brain is not simply a machine operating on auto pilot, but rather a dynamic, ever changing, and adaptable system. It's about recognizing that there are scientific processes at work, and through that very understanding we may begin to build a path to recovery. This next section will explore how that recovery might begin, and how these neural systems can, over time, be brought back into a healthy state of equilibrium.

Neuroscience of Pornography Addiction

When Protection Goes Too Far

Having explored the initial sparks of desire, the establishment of reward pathways, and the dangerous terrain of excitotoxicity, we now arrive at a particularly intricate stage of the addiction process: the point at which the brain's own defense mechanisms, designed to protect it from harm, inadvertently contribute to the deepening of the cycle. This is a sort of neurological irony, where the very systems meant to safeguard us from overstimulation can ultimately create conditions that perpetuate and even exacerbate the addiction. It is as if the body's defenses are working too hard, and inadvertently creating a bigger problem than they intended to solve, and we will be diving deep into that very complex system, that operates like a set of fail safes that have gone horribly wrong. The brain's resilience is remarkable, its ability to adapt is astounding, but even the most finely tuned systems can, when pushed to their limits, begin to create new problems in their attempt to solve old ones.

The primary player in this phase of the cycle is a fascinating molecule known as Delta FosB. Delta FosB is not a neurotransmitter in the traditional sense, but a transcription factor, a protein that regulates gene expression. It's like a master switch that turns specific genes on or off within a cell, thereby controlling the production of other proteins that alter neural circuitry. It doesn't act immediately like a neurotransmitter, but rather, works more slowly, creating long lasting changes in the brain's architecture. In the context of addiction, Delta FosB is activated by repeated exposure to addictive substances or behaviors, such as viewing pornography. The research of Eric Nestler, particularly his 2008 paper, "Transcriptional mechanisms of addiction: the role of DeltaFosB" has been pivotal in illuminating the role of Delta FosB in the neurobiology of addiction. Nestler details how Delta FosB accumulates in the brain as it experiences repeated stimulation, creating long lasting alterations to gene expression that ultimately contribute to the development of addiction. Think of it like a signal that the brain sends when it knows it has to start making some lasting changes. Initially, the presence of Delta FosB may seem beneficial, working to protect neurons from the potentially damaging effects of excitotoxicity, reducing some of the sensitivity in the dopamine receptors, but, in the long run, this very protective mechanism contributes to the development of tolerance, the very phenomenon where the individual becomes less sensitive to the stimuli that once provided pleasure.

This brings us to another key player: Brain-Derived Neurotrophic Factor, or BDNF. BDNF is a neurotrophin, a protein that supports the survival, growth, and differentiation of neurons. It is also crucial for neural plasticity, the brain's ability to form new connections and adapt to new experiences. While BDNF plays a critical role in learning and memory, in the context of addiction, it is implicated in a very specific type of neural adaptation. When Delta FosB is activated it triggers the production of BDNF, particularly in the reward areas of the brain. But, rather than promoting a healthy adaptation, the BDNF, in this context, begins to strengthen the neural connections associated with the addictive behavior, and in so doing, makes it harder to break free. It begins to act like the fortifications around a castle, strengthening the walls and making it increasingly challenging to breach the fortress of the pleasure circuitry. It's as if the brain, recognizing that it is being repeatedly bombarded with stimulation, begins to build up defenses, creating new neural pathways and reinforcing the existing ones so that that behavior, and the desire for it, becomes more deeply ingrained and ingrained into the brain. In this way, BDNF, while essential for brain health, contributes to the entrenchment of addiction by making the very behaviors that are the root of the problem become more difficult to change. It's like adding more concrete to an already established wall, reinforcing the structure, but not necessarily making it a healthy or sustainable structure.

Neuroscience of Pornography Addiction

What is the ultimate outcome of these protective measures? One of the most significant is the intensification of desire, and a deepening of the cycle of dependence. As the brain becomes less sensitive to the effects of dopamine due to the presence of Delta FosB, and it starts to reinforce its own reward pathways through BDNF, the individual will require increasingly stronger or more varied forms of stimulation to achieve the same level of pleasure or satisfaction. This is the phenomenon known as tolerance, and it leads to the brain needing more and more, to reach the desired state. This isn't simply about a desire for more pleasure, it's a response to the brain's diminishing capacity to experience pleasure from its normal stimulations. The brain, in its attempt to protect itself from the harmful effects of overstimulation, inadvertently sets the stage for an intensification of the addictive behavior. It's as though the body is saying, "I'm not getting what I need, so I'm going to ramp up the demand." This is not only a physical shift, but a psychological one as well, where the focus of the individual becomes more and more centered around the addictive behavior, creating an unending cycle of increased demand and diminished satisfaction. It's like turning the volume up on a stereo system that has started to distort, in an attempt to get more clarity and definition, all the while the system is being pushed further and further beyond its intended parameters.

And beyond the concept of tolerance, there's another equally important effect: the dysregulation of the reward system. The reward system, as we know, is not just about experiencing pleasure; it's also about motivation, desire, and the drive to pursue goals. But, when the brain becomes dysregulated, that reward system starts to function in an erratic and unpredictable way. The natural cues of pleasure and reward become distorted, and they are all but hijacked by the addictive behavior. The brain starts to respond more strongly to the artificial stimuli associated with pornography than it does to the everyday experiences that once brought joy or satisfaction. The natural ability to feel pleasure, to feel motivated, and to feel joy becomes compromised by the altered reward circuitry. This is not simply a matter of wanting more pornography; it's about the brain's inability to respond to other sources of reward. It's like having a favorite dish on the table, but then suddenly having that one dish become the only thing you can taste. The other flavors have, essentially, disappeared, and the individual starts to experience the other things in their life as dull, bland or uninteresting. The result is a narrowing of one's focus, with everything else being diminished in importance compared to the addictive behavior. This is how the addictive behavior becomes the central and consuming focus in the life of an addicted individual, while all other aspects of their life begin to diminish.

The final piece of this complex puzzle is what has been described as the "cycle of reinforcement." We've already discussed how the body tries to create balance, but when those balances are pushed too far, the very defensive mechanisms that the body has developed for self-preservation create their own problems. The body, through Delta FosB and BDNF, is trying to protect itself from the harmful effects of excessive stimulation, but those very adaptations, paradoxically, make the addictive behavior more deeply ingrained. They make it harder to break free from the cycle, and they make it so that the addictive behavior is now the only thing that can stimulate that pleasure circuitry. The brain learns that it needs the addictive behavior to avoid the dissatisfaction created by the brain's own defense mechanisms. It is now a situation where the brain has, in essence, become trapped by its own devices. It's like a labyrinth that's designed to lead you out, but ultimately ends up leading you deeper inside, and it is this that creates an incredibly difficult challenge for the individual attempting to break free from the cycle of addiction.

The brain's defense mechanisms, represented by Delta FosB and BDNF, are not always our allies, and in the context of addiction, these protective systems often contribute to the perpetuation of the addiction. Tolerance develops, desire intensifies, the reward system becomes dysregulated, and a complex cycle of reinforcement is put into motion. This underscores the fact that addiction is not simply a lack of willpower, but the result of deep-seated neurological processes that

Neuroscience of Pornography Addiction

can, if left unchecked, create a feedback loop that is very difficult to escape. The brain's very attempt to regulate itself has led to a rewiring that makes the addictive behavior the central focus, and this leads to a system that is very hard to disengage from. By understanding these complex mechanisms, we can begin to see that this is not a simple matter of choice or desire, but a complex interplay of brain changes, defense systems, and biological responses. This understanding is crucial as we move forward to explore the path to recovery, a journey that involves not only reversing the neurological changes, but also retraining the brain to respond to healthier and more fulfilling experiences. The path to breaking free from this cycle is not simple, but by understanding these complex systems, we begin to see that change, transformation, and healing, are still possible.

Kendir Ramiz

Addiction and its Consequences

Having explored the brain's defense mechanisms and their paradoxical role in perpetuating the addiction cycle, we now turn our attention to the stark realities of the downward spiral. This is the point where the consequences of chronic pornography use become increasingly evident, impacting not only the brain's reward system but also extending to various aspects of physical, mental, and emotional health. It's like watching a complex and elaborate structure gradually collapse in on itself, the foundational cracks now beginning to spread to every part of the system. This is not simply a matter of increased tolerance or a more intense craving; this is a deep systemic breakdown, a fundamental dysregulation that manifests in many deeply concerning ways. It's where the abstract neurochemical processes we've explored become deeply personal and often painful realities. The scientific terms that we have

Neuroscience of Pornography Addiction

explored are not just scientific terms, they are representative of very real struggles that people go through, and it is time to make those processes far more visible.

One of the most significant consequences of chronic pornography use is a profound reduction in the brain's sensitivity to dopamine. We've discussed dopamine's role as a key neurotransmitter in the reward system, the chemical messenger that drives motivation and pleasure. But when the brain is repeatedly flooded with high levels of dopamine due to pornography use, the dopamine receptors begin to downregulate, becoming less responsive to normal levels of this crucial neurotransmitter. It's as if the brain is adapting to the constant bombardment by turning down the volume, or by becoming less sensitive to the incoming signals. This is not simply a matter of reduced sexual pleasure; it is an overall diminishment in the capacity to experience pleasure from any activity, even those that once brought joy and satisfaction. The things that once provided enjoyment, from spending time with loved ones to accomplishing goals, become less rewarding, less compelling. The world around starts to feel muted, and the individual starts to lose the ability to experience life with its full range of sensations, and this reduction in dopamine sensitivity begins to impact almost all other aspects of life. This is not merely a psychological phenomenon; it's a neurological change, a physical alteration in how the brain responds to the environment. The brain, in a sense, has become dulled to what once provided it with pleasure.

This reduction in dopamine sensitivity leads to what many describe as a state of "flatlining" or apathy. The world begins to lose its color, its vibrancy, its capacity to elicit emotional responses. This isn't simply a feeling of sadness or boredom; it's a deeper sense of detachment, a lack of motivation, and an inability to experience the natural joys of life. It's as though a dimmer switch has been turned down on life itself, muting all of the sensations. Simple tasks, like getting out of bed, interacting with others, or engaging in hobbies, become increasingly challenging, requiring far more effort than they used to. This is not a matter of laziness or a lack of willpower; it's a manifestation of the brain's reduced capacity to experience reward and motivation. This lack of dopamine does not simply affect the emotional states, but impacts cognitive abilities as well. Many individuals report experiencing "brain fog," a difficulty in thinking clearly, concentrating, and remembering things. The mental landscape, once vibrant and sharp, becomes murky and clouded, and the individual begins to struggle with simple cognitive tasks. This is not simply a temporary state; it can be a pervasive and persistent experience that impacts every area of life, further contributing to feelings of frustration, hopelessness, and isolation. The reduction of dopamine levels, and the resultant impact on our moods and cognitive states, is not a subtle effect, but it is profound.

Concurrently, another deeply troubling shift occurs in the brain's stress response system. The HPA axis, the same system that initially releases cortisol to motivate the pursuit of desire, becomes chronically activated. This is like having the body's emergency alarm system constantly going off, even when there is no immediate threat. In the previous chapter, we discussed how the HPA system kicks into gear, and with this chronic activation, it remains in a hypervigilant and high alert state. The constant demand for dopamine, driven by the addiction, and the resultant reduction in dopamine's effectiveness, places significant stress on the brain. This leads to an elevated level of stress hormones, like cortisol, circulating in the system, and this is not a temporary state. Over time, this chronic activation of the stress response has profound consequences for physical, mental, and emotional well-being. Individuals may experience persistent anxiety, irritability, restlessness, and an inability to relax. This isn't just a matter of feeling slightly more on edge, this is a constant state of unease, and an increased sense of nervousness, with very little relief. Sleep patterns are often disrupted, leading to chronic insomnia, and fatigue. This lack of sleep further exacerbates the already existing mood disturbances and cognitive impairments, creating a vicious cycle that is incredibly difficult to break. This chronic activation of the HPA axis

Neuroscience of Pornography Addiction

is not just unpleasant; it's a significant burden on the body and mind, further contributing to the overall deterioration of health. The body is in a state of perpetual fight or flight, which is not conducive to long term health, or happiness.

This combination of reduced dopamine sensitivity and chronic stress activation creates the perfect storm for an intensification of cravings. Cravings are more than just an intense desire for pornography, they are a neurobiological response, driven by complex interactions between the reward and stress systems. When dopamine levels drop, and stress hormones spike, the brain signals an intense need for the very thing it has become addicted to, in an attempt to reestablish a sense of balance. It's a desperate attempt to alleviate the discomfort caused by the dysregulation of the neurochemical system. These cravings can be incredibly difficult to resist, often overpowering rational thought and conscious intention. It's not just a matter of a strong desire; it's a physiological response, like a powerful urge that takes over the mind and the body.

Kendir Ramiz

The intensity of these cravings can be so severe that individuals may feel as though they have no choice but to engage in the addictive behavior, even when they understand it to be harmful. It's like an itch that must be scratched, no matter what the consequences, and this cycle of craving, and engagement further exacerbates all of the other issues, creating a deep feedback loop that makes it harder and harder to escape.

Beyond the direct effects on the brain, chronic pornography use also has significant impacts on physical and mental health. The constant activation of the stress response, combined with poor sleep, and poor nutrition, can manifest as a variety of physical symptoms, from chronic fatigue to gastrointestinal issues. The emotional toll can be even more profound. Chronic pornography use is often associated with increased rates of anxiety, depression, and low self-esteem. It can also negatively impact body image and self-perception, with individuals often feeling inadequate or ashamed, especially when comparing themselves to the idealized images presented in

Neuroscience of Pornography Addiction

pornography. This is not simply a byproduct of the addiction, it is often a result of the altered brain chemistry as well, as the neurotransmitters that are responsible for mood regulation are impacted by the addiction cycle. As explored in reports from the National Institute on Drug Abuse, these negative mental health consequences can range from mild to severe, contributing to a sense of isolation, hopelessness, and at times even suicidal ideation. This is not merely a matter of psychological distress; it is often the result of real physical changes in brain function. This demonstrates that the impacts of addiction do not remain within the brain, but manifest physically and emotionally, often creating a cascade of negative consequences.

The downward spiral of addiction is not simply a matter of willpower or a lack of self-control; it's a complex neurobiological process, that can result in a cascade of negative consequences. Reduced dopamine sensitivity, chronic stress activation, and intense cravings, are not just psychological symptoms but, are the manifestations of very real neurochemical changes in the brain. These changes are not just limited to the brain but extend to overall physical and mental health, often resulting in a devastating impact on many areas of life. Recognizing the complex and pervasive consequences of chronic pornography use is the first step in understanding how to approach recovery. This underscores the importance of addressing not only the immediate addictive behavior, but also the underlying neurological and psychological factors that contribute to this complex problem. The journey towards recovery is not going to be a simple one, but armed with this knowledge, we can begin to approach this challenge with a deeper sense of compassion, wisdom, and a greater understanding of what is actually happening. And while the consequences are severe, the capacity of the brain to heal, transform, and recover, still remains, and that is the seed of hope as we move forward into the next chapter.

Strategies for Recovery

Having navigated the intricacies of the downward spiral, it is now time to turn our attention towards the beacon of hope: the strategies for recovery. This chapter is dedicated to exploring the essential steps involved in reversing the neurological changes, retraining the reward pathways, and reclaiming a life that has been deeply affected by addiction. This isn't simply a matter of "stopping," but rather, a journey of transformation and healing, a process that requires a multifaceted approach and a deep understanding of the underlying neurochemical processes. The path to recovery is not a simple one, but it is always possible. And in this chapter we will begin to explore the practical steps involved in that challenging but ultimately liberating journey. It is important to recognize that there is no magic cure, no simple solution; it is a process that requires dedication, patience, and a deep commitment to oneself.

The first, and perhaps most challenging step on this journey is the necessity of abstinence. This is not simply a matter of avoiding pornography, it's a fundamental step in allowing the brain to begin its repair process. Abstinence is crucial for reducing the levels of excitatory neurotransmitters, such as glutamate and dopamine, that have become chronically elevated due to addiction. It is like removing the fuel source from a fire, that will allow that fire to eventually die out. When the brain is constantly bombarded with the intense stimuli of pornography, it becomes locked in a perpetual cycle of stimulation and reward, and the brain needs a break from that cycle in order to begin the healing process. Abstinence allows the brain to reset, to begin to reestablish a more natural balance in the levels of neurotransmitters and hormones. It is the starting point for restoring healthy brain function and allowing the natural processes of regulation to kick back in, to allow the overstimulated brain to find its natural state. This is by no means an easy process, but it is a crucial component of the recovery journey. It is the first step on

Neuroscience of Pornography Addiction

a path towards healing, a declaration of intent that one is choosing a different and healthier life.

The path of abstinence is rarely a smooth one, and it is often characterized by the challenging experience of withdrawal. When the brain becomes accustomed to a certain level of stimulation, abruptly stopping that stimulation can trigger a range of physiological and psychological symptoms. These withdrawal symptoms can be intensely uncomfortable, even debilitating. Intense cravings, irritability, anxiety, and hypersensitivity to stimuli are common experiences. The cravings can feel all consuming, like a physical ache that can only be relieved by the addictive behavior, and these cravings are not simply a matter of wanting pornography, it is an actual neurological response, as the brain screams out for the neurochemicals that it has become accustomed to. Irritability, is not simply a matter of being in a bad mood, it's a manifestation of the stress response system as it struggles to regain its balance, and the anxiety and hypersensitivity to stimuli are not simply psychological, but they are the result of a dysregulated stress response and changes to the neurological system. These symptoms can vary in intensity, and for some it may be a subtle shift in their mood, and for others it can be incredibly overwhelming, and sometimes, even dangerous. What is important to understand is that this is not simply a matter of weakness, but it is a very real neurochemical process. Navigating the challenges of withdrawal requires a great deal of self-compassion, a good support system, and an understanding that these symptoms are temporary and an integral part of the recovery process. The withdrawal process is not simply something to be endured, but rather it is a sign that the body is beginning to heal, that it is beginning to adjust to a life without the addictive behavior. It is important to remember that these symptoms are temporary, and they represent a sign that change is on the horizon.

But, here is some good news: the brain is not a static structure; it is a dynamic and ever-changing system, capable of remarkable adaptation through neuroplasticity. Neuroplasticity refers to the brain's ability to reorganize itself by forming new neural

Neuroscience of Pornography Addiction

connections throughout life. It is the very basis of learning and adaptation, and it is also the foundation of all healing. This means that the negative changes in brain structure and function associated with addiction are not necessarily permanent, they can be reversed, they can be transformed. As we start the process of abstaining from the addictive behavior, this neuroplasticity begins to kick in and begins the process of rewiring, and reorganizing the brain's neural pathways. This offers a great sense of hope, for it means that even after years of chronic addiction, the brain has the capacity to heal and to rebuild itself. The challenge is not merely about enduring the discomfort of withdrawal, but to actively support the brain's natural ability to rewire, and to build healthier neural pathways. This is not something that happens automatically, it requires active engagement, commitment and consistent effort. This is the good news, the possibility for transformation and the hope for a better future, if one is willing to engage with the process.

Central to this process of rewiring is the reconditioning of the brain's reward system. As we've discussed previously, the brain's reward system has been hijacked by the intense stimuli of pornography, and this must be intentionally re-trained to respond to healthier sources of pleasure. This is not simply about avoiding pornography, it's about developing new neural pathways that connect the experience of pleasure, joy and satisfaction with healthy activities. There are a variety of therapeutic techniques that can help to recondition the reward system, including mindfulness practices, cognitive behavioral therapy (CBT), and other forms of talk therapy. Mindfulness practices can help individuals become more aware of their thoughts, feelings, and cravings without judgment, allowing them to observe their impulses and make more conscious choices about how to respond. CBT is an evidence based approach that focuses on identifying and changing negative thought patterns and behaviors, and it can be particularly effective in helping individuals develop healthier coping strategies for managing cravings and avoiding triggers. Talk therapy provides an opportunity for individuals to explore the root causes of their addiction, often uncovering traumas or emotional issues that may be contributing to the cycle of addictive behaviors. These different therapeutic techniques do not necessarily work independently, they all work together to help an individual move through the recovery process and rewire their brain. This process of reconditioning is not a quick fix; it's a gradual process of retraining the brain, a process that requires patience, persistence, and self-compassion. It is a process of consciously building new associations, new patterns of behavior, and new responses to the old triggers.

Lastly, and equally important, is the role of therapeutic interventions and support systems. Recovery is not something that can often be achieved alone; it often requires the guidance and support of trained professionals and others who are on similar journeys. As detailed in the various publications of the American Psychological Association, the use of different therapies, support groups, and a comprehensive treatment plan have been proven to increase the likelihood of a successful recovery. Therapeutic interventions, such as individual therapy, couples therapy, or group therapy, provide individuals with a safe and supportive space to explore their addiction, to develop healthier coping mechanisms, and to build stronger connections to others. Support groups, such as 12-step programs, provide a sense of community, and reduce the sense of isolation, offering a place where individuals can share their struggles and their successes with others who truly understand what they are going through. A comprehensive treatment plan is also key, as it addresses all the various aspects of the individual, including physical, mental, and emotional health, in a unified approach. This is not just about treating the symptoms of addiction; it's about treating the whole person, recognizing the complex interplay of biological, psychological, and social factors that contribute to the addictive cycle. The use of therapeutic interventions and support systems is not a sign of weakness, it is a clear indication that one understands the depth of the problem, and is committed to using all of the tools available to create a healthier and more fulfilling life.

Neuroscience of Pornography Addiction

Reversing the tide of addiction is a complex process that requires commitment, patience, and a deep understanding of the underlying neurochemical mechanisms. Abstinence is the crucial first step, allowing the brain to begin its healing process. Navigating withdrawal symptoms requires self-compassion and support. The brain's remarkable capacity for neuroplasticity offers hope for lasting transformation. Reconditioning the reward system involves training the brain to respond to healthier sources of pleasure and therapeutic interventions and support systems are essential for providing a space to explore the root causes of addiction. By understanding these complexities and implementing a multi-faceted approach that honors both the biological and psychological factors, individuals can begin to reclaim their lives, rebuild their brains, and create a more sustainable and fulfilling life. The capacity for healing is always present, and with a combination of self-awareness, dedication and support, lasting transformation is absolutely possible.

Kendir Ramiz

The Mind-Body Connection in Recovery

Having explored the neurological and psychological strategies for recovery, we must now turn our attention to a fundamental and often overlooked aspect of the healing process: the inextricable link between the mind and body. This chapter is dedicated to exploring the holistic approach, recognizing that the journey to recovery is not solely a matter of rewiring the brain or addressing psychological patterns, but also a process that involves nurturing the physical body, creating a supportive and conducive environment for lasting transformation. It is as though we are recognizing that the mind and body are not separate entities, but rather, two parts of a unified whole, each intimately connected to the other, each influencing the other in a continuous dance of give and take. This is not merely a conceptual idea; it's a practical and scientifically grounded approach

Neuroscience of Pornography Addiction

that acknowledges the deep interconnectedness of our physical, mental, and emotional states. The recovery journey must honor that deeply integrated system, understanding that healing can't simply be something that is intellectual, it must also be something that is deeply experiential, and embodied.

Let's start with sleep, often one of the most neglected pillars of health, but particularly crucial for the brain's repair process. When we sleep, our brain does not simply switch off; it enters a highly active state of repair and consolidation. During sleep, the brain clears out toxins, consolidates memories, and allows various neural circuits to restore themselves after the day's activities. Lack of sleep, or poor sleep quality, disrupts these essential processes, making it far more difficult for the brain to heal, to rewire, and to adapt. Chronic sleep deprivation can exacerbate the symptoms of withdrawal, making cravings more intense and reducing the capacity to regulate emotions, and make clear decisions. In the context of recovery, prioritizing sleep, is not simply a matter of getting enough rest, it is about creating the ideal conditions for the brain to do the important work of repair. This is why setting a regular sleep schedule, creating a peaceful sleep environment, and developing healthy sleep hygiene practices are all important components of a comprehensive recovery plan. It's about understanding that sleep is not a luxury; it's a fundamental biological necessity that supports all aspects of physical and mental health. It is at this point where the body and mind begin to heal in a synchronized manner, and poor sleep directly impacts both.

Next, let's turn our attention to nutrition and its critical role in brain health. The food we eat is not just fuel for our bodies; it's also the building blocks for our brains. Proper nutrition provides the essential nutrients that support neuronal function, neurotransmitter production, and the overall health of the gut-brain axis. A diet rich in whole foods, fruits, vegetables, lean proteins, and healthy fats is crucial for restoring the brain's balance and supporting the production of the necessary neurotransmitters involved in mood regulation, motivation, and cognitive function. But beyond the individual food groups, it is also important to acknowledge the role of the gut microbiome, the community of microorganisms that live in our digestive system. The gut and brain are in a constant state of communication and poor gut health can impact brain function, mood, and overall health, through that communication. This complex relationship, known as the gut-brain axis, highlights that what happens in our digestive system has a profound impact on our mental health. In the context of recovery, nourishing our bodies with healthy foods that support gut health is not just about physical wellness; it is about creating a supportive foundation for a healthy mind. This is not a merely a suggestion, it is a fundamental aspect of the recovery journey, and choosing foods that are nutritious are a sign that one has decided to value themselves, and in so doing, values their own healing. It's not simply about avoiding unhealthy foods, it's about consciously choosing foods that provide the brain with the essential nutrients it needs to restore its function.

And beyond just the food, we also must acknowledge the role of physical exercise as a powerful tool in recovery, and not simply as an additional component of health. Regular physical exercise is not simply a way to burn calories or build muscles; it's a powerful way to promote the production of neurochemicals, and improve cognitive function, reduce stress and promote overall health. Exercise stimulates the release of endorphins, the body's natural mood boosters, which can improve feelings of well-being, and reduce stress and anxiety. Exercise also increases the production of BDNF, a neurotrophin that we explored in the previous chapters that is crucial for neuroplasticity, the brain's capacity to rewire itself. Regular physical activity, can also improve sleep, increase energy levels, and reduce the symptoms of depression and anxiety. It's as though the body and mind respond in a coordinated manner to the physical exercise, with all of the systems being brought back into balance. In the context of recovery, exercise is not simply a supplementary activity; it's a valuable component of a comprehensive approach to healing and transformation. It's a way of consciously building new habits, new routines, and new neural pathways, replacing the old patterns with ones that support health, joy and overall well-being. It's not merely a physical activity, it is a form of active self-care.

The practice of mindfulness and stress reduction is another key aspect of the mind-body connection in recovery. Mindfulness, in this context, is about cultivating a non-judgmental awareness of the present moment, and being present to what is happening right here and now, rather than dwelling in the past, or worrying about the future. Through practices like meditation, deep breathing exercises, and mindful movement, individuals can learn to observe their thoughts and feelings without being overwhelmed by them. Mindfulness allows us to create a buffer between stimulus and response, providing space to choose how we react, rather than being ruled by impulsive urges. It is a skill, and like any skill it requires consistent effort to develop, but the payoff is increased self-awareness, improved emotional regulation and reduced stress. In the context of recovery, mindfulness practices can be invaluable tools for managing cravings, reducing anxiety, and cultivating a sense of inner peace, offering a way to manage all of the various symptoms in a way that does not require the addictive behavior. It's not simply about avoiding negative emotions; it's about creating a space for observing them without judgment, for acknowledging them without being overwhelmed by them.

Let us acknowledge the importance of taking a holistic approach to recovery. This is not about focusing solely on either the mind or the body but about recognizing the intricate and reciprocal relationship between the two. The mind and body are not isolated systems; they are interconnected, and they influence each other in a constant, ongoing process. When we address the physical needs of the body through proper sleep, nutrition, and exercise, we are also supporting the health of the mind. And, when we nurture the mind through mindfulness and stress reduction practices, we are also supporting the overall health of the body. This is not a compartmentalized approach; it's an integrated way of understanding health and well-being. In the

Neuroscience of Pornography Addiction

context of recovery, a holistic approach means addressing all aspects of the individual—physical, mental, emotional, and spiritual—creating a foundation for lasting transformation. It is a recognition that health is not simply an absence of disease; it's an active process of self-care, self-awareness, and self-compassion. This also acknowledges that recovery is not something that can be accomplished quickly, it requires time, effort, commitment and a deep understanding of the interconnected nature of the body and the mind.

The mind-body connection is a vital aspect of the recovery journey, and should not be overlooked. Adequate, high quality sleep is critical for brain repair. Proper nutrition supports neuronal function and gut health. Regular physical exercise promotes the production of neurochemicals and improves overall well-being. Mindfulness practices reduce stress and improve emotional regulation and, finally, a holistic approach recognizes the intricate relationship between the mind and body. Addressing both the physical and mental aspects of addiction is essential for creating a comprehensive approach to recovery that honors the deep connection of all aspects of the human experience. The journey to healing is not simply about breaking free from an addiction; it's about building a life that is rooted in self-care, self-awareness, and self-compassion, and by respecting this deeply integrated system, we may find a path to lasting health, peace, and well-being. It's not simply a matter of healing, it's about cultivating a more conscious and connected relationship with our own bodies and minds. This is not just a part of recovery, it is a path towards a richer, more fulfilling way of being.

Building Meaningful Connections

Having explored the intricate pathways of the brain, the physiological mechanisms of recovery, and the essential role of the mind-body connection, we now turn our attention to another vital component of healing: the power of human connection. This chapter is dedicated to exploring the crucial role of meaningful relationships in recovery, recognizing that addiction is not just an individual struggle, but often a deeply isolating experience. It is about acknowledging that human beings are fundamentally social creatures, wired for connection and belonging, and that these very connections can be powerful forces in fostering resilience, healing, and growth. It's as though we are acknowledging that we do not exist in a vacuum, that our lives are inextricably interwoven with the lives of others, and these relationships often become the very foundation upon which our recovery is built. This is more

Kendir Ramiz

than just a simple concept, it's about the fundamental need to be seen, heard and known for who we are, not just in our own minds, but in the eyes of others, and in those relationships we may find the space to explore who we are without the layers of addiction.

Let's begin with the fundamental importance of social connection in the recovery process. It is not simply that we like to be around other people, it is that we need it. Healthy social connections are not merely a luxury; they are a basic human need, as essential to our well-being as food, water, and sleep. When we are socially connected, we experience a sense of belonging, of being seen, and understood, and valued for our inherent worth. These connections provide us with a sense of purpose, a sense of meaning, and a sense of support during times of challenge and difficulty, which is often the case in any recovery process. Conversely, isolation, and loneliness can have a profoundly negative impact on mental and physical health, increasing the risk of depression, anxiety, and a whole range of other physical ailments. In the context of addiction, social isolation

Neuroscience of Pornography Addiction

often exacerbates the cycle of compulsive behavior, creating a feedback loop of shame, secrecy, and further isolation. This is not simply a matter of being alone, it is a state of disconnection, and when someone is disconnected they are far more likely to fall further into the patterns of addiction. Recovery from addiction often requires a conscious effort to break free from this cycle, and to actively build healthier social connections that can foster healing, growth, and a sense of belonging. We must acknowledge that healing often happens in community, not in isolation, and that support system is often the cornerstone to the long term healing process.

Not all connections are created equal, and the quality of our relationships matters deeply, and this is where the power of vulnerability and intimacy comes into play. Vulnerability, is often seen as weakness, but it is actually one of the greatest strengths, and it is the gateway to intimacy and the truest form of human connection. It's about having the courage to show up as your authentic self, flaws and all, to share your deepest fears, your greatest hopes, and your most vulnerable parts, and to not have to be perfect in order to be loved and accepted. In the context of recovery, vulnerability is not simply an emotional act, it's a transformative process that allows individuals to break free from the shame and secrecy that often accompany addiction, and in so doing, build the authentic connections that are so crucial to the healing process. Intimacy, is not just about romantic relationships, it is about the ability to connect with another person on a deep and meaningful level, to feel truly seen, known, and accepted. It is in these spaces of genuine vulnerability, that genuine connection is built. These types of relationships provide a sense of safety and support, where individuals can explore their feelings, process their experiences, and rebuild their sense of self. This is not about being perfect, or not making any mistakes; it is about accepting ourselves, exactly as we are, and in so doing, creating the environment where we can be seen, and heard, for who we truly are. The opposite of addiction is not sobriety, the opposite of addiction is connection, and it is in these authentic, vulnerable connections that we may find true freedom.

Neuroscience of Pornography Addiction

And, as a way to begin to foster those authentic relationships, it is important to actively develop a robust support network. This goes beyond our immediate friends and family, and extends to building a community of support, that can help to navigate the challenges of recovery. These types of support systems often take the form of peer groups and support groups, spaces where individuals can connect with others who have also experienced addiction. Peer groups, provide a space for shared understanding and validation, as well as create an environment where individuals can share their experiences, their challenges, and their successes without judgment. The sense of community that these types of peer groups provide can help to reduce the feelings of isolation, shame, and guilt that often accompany addiction, and they create the feeling of being seen, known, understood, and not alone. These groups are not simply a place to share stories; they are also a source of accountability and motivation, as members support each other along the recovery journey. Support groups, often guided by trained professionals, provide a structured and therapeutic setting for addressing the underlying issues that contribute to addiction, and for learning new coping skills, and strategies to maintain long-term sobriety. These support systems are not a sign of weakness, they are a sign of strength, of understanding the need for community, and the need to not navigate this journey alone.

We must also recognize the importance of healing relationships, not just building new ones. Addiction often takes a toll on existing relationships, damaging trust, and creating rifts with loved ones. Repairing these damaged connections is a critical step in the recovery process, and it often requires patience, humility, and a willingness to acknowledge the impact of one's actions. This often involves sincere apologies, open and honest communication, and a willingness to make amends for past behaviors. This process of healing is not easy, but it is an essential component of restoring a sense of belonging, reconnecting to others and rebuilding damaged trust. When relationships are healed, they can become a strong source of support, providing love, compassion, and a sense of belonging, and in so doing, they also provide the space for the individual to heal as well. The healing is not a one-way street, it is a process where everyone involved in those relationships, has the chance to heal, to grow, and to learn. This process is not simply about returning to the past; it's about creating a new future where relationships are built on a foundation of authenticity, trust, and mutual respect.

Explore the power of authenticity and self-expression. Addiction often forces individuals to hide parts of themselves, to suppress their feelings, and to live in a constant state of fear of being exposed. In the process of recovery, it becomes essential to reclaim one's voice, to express thoughts and feelings openly, and to allow oneself to be seen for who they truly are. Authenticity, is about being true to oneself, living in accordance with one's values, and allowing one's inner self to be seen by the world. Self-expression, is about finding healthy and creative ways to communicate one's thoughts, feelings, and desires. These could take the form of art, music, writing, movement, or any other activity that allows for genuine self-expression. When individuals are able to live authentically, and to express themselves openly, they create a stronger sense of self, a more integrated and congruent identity, and can build deeper, more meaningful connections with others. This is not simply a matter of personal preference, it's a vital step in reclaiming one's life, and in choosing to not live a life based on a lie.

Building meaningful connections is a vital part of the recovery journey, not simply a helpful add-on. Social connections foster belonging and support. Vulnerability and intimacy allow for authentic relationships. Developing a support network reduces isolation, and provides a sense of community. Healing relationships restores trust and provides love and compassion. And, authenticity and self-expression empower individuals to live true to themselves and build authentic connections. The journey from addiction to recovery is not simply an individual one, it is often a collective one, requiring the support of others and a deep commitment to building relationships that provide safety, love, understanding and belonging. By prioritizing these human connections, we can create a foundation for lasting healing, growth, and a more fulfilling life, and to move beyond the shadows of addiction and step into the light of meaningful and authentic connection. The path to recovery is one that is often best walked together, and in that togetherness we often find the strength, and the hope, to move forward. It's not simply about healing the individual, it is also about healing the collective, and in that collective healing, we begin to heal ourselves.

Kendir Ramiz

Reframing Pleasure and Desire

Having explored the crucial role of social connections in recovery, we now turn our attention inward, to the very heart of our experience: the nature of pleasure and desire. This chapter is dedicated to exploring the ways in which we can reframe our relationship with pleasure, moving beyond the instant gratification of hyper-stimulation and embracing a more nuanced and meaningful experience of satisfaction. It is about understanding that pleasure is not simply a fleeting sensation, but a complex and multifaceted experience that is deeply intertwined with our values, our goals, and our overall sense of well-being. It's as though we are becoming the architects of our own desires, intentionally choosing the things that are truly satisfying, rather than simply responding to the cravings and impulses that

Neuroscience of Pornography Addiction

have led to so much difficulty. This is a process of conscious re-evaluation, and by choosing to engage with pleasure more intentionally, we move out of the cycle of addiction, and move into a space of intentional living. It's not about denying ourselves pleasure, but about choosing a more sustainable, fulfilling, and ultimately, more meaningful expression of it.

Let's begin with the crucial step of moving beyond instant gratification. As we have discussed at length throughout this book, the brain's reward system has been hijacked by the intense and readily available stimuli of pornography, creating a cycle where the brain is constantly seeking a quick dopamine hit, often at the expense of longer-term satisfaction. In recovery, it is essential to intentionally explore other sources of pleasure, those that lie outside of the realm of hyper-stimulation. This requires a conscious effort to discover activities that bring genuine joy and contentment, not simply a momentary surge of dopamine. These might include spending time in nature, engaging in creative pursuits, cultivating meaningful relationships, learning new skills, or participating in activities that align with one's values. This is not about finding activities that are more "acceptable" than pornography, but more about building a life that is filled with a wide variety of satisfying experiences, those that offer a more sustainable and integrated type of pleasure. It's as if we are choosing to savor a gourmet meal, rather than simply grabbing a quick and unhealthy snack. This is a journey of rediscovery, a process of learning what truly satisfies us, and to make conscious choices to choose those experiences. This is not simply about avoiding the old addictive behaviors, it's about actively choosing behaviors that are supportive of long term health, and well being.

Discuss the immense value of effort and achievement in the pursuit of genuine satisfaction. The brain is not just wired for immediate gratification, but for the satisfaction that comes with working towards a goal, and in seeing those goals come to fruition. In a world where instant gratification is so easily available, it is easy to overlook the deep sense of fulfillment that can come from effort,

Neuroscience of Pornography Addiction

perseverance and dedication. Working towards a longer-term goal, whether that is a creative project, a fitness goal, a career aspiration, or any other meaningful pursuit, can create a far more sustainable sense of satisfaction than the fleeting pleasure of a quick dopamine hit. The satisfaction that is derived from achievement is not simply a dopamine release; it is a deeper sense of self-efficacy, of competence, and a sense of mastery that comes from the process itself. This is about understanding that true pleasure is not about the immediate outcome, it is also about the process of getting there. This type of pleasure and satisfaction is not simply a substitute for the old patterns, it's an entirely different type of engagement with life itself. The world is filled with opportunities to grow, to learn, to create and to achieve, and these opportunities are often far more rewarding than seeking the easy path of instant gratification.

This brings us to the importance of mindful engagement in our daily activities, whatever those activities may be. Mindful engagement is about bringing our full

Kendir Ramiz

awareness to whatever we are doing, focusing on the present moment, rather than simply going through the motions. It means paying attention to the sensory details of our experience, noticing the small things, and finding the joy in the everyday moments. This could include eating a meal with full awareness of the taste and texture of the food, listening intently to a piece of music, taking a walk in nature and appreciating the beauty, or simply being present in a conversation with another person. When we are engaged in this way we create a deeper connection to ourselves, our experiences, and the world around us. Mindful engagement is not simply about focusing our attention, it's about engaging with our lives in a way that is authentic, intentional, and deeply enriching. It allows us to find pleasure in the simple things, to experience the beauty of life in its many forms, and it is a powerful antidote to the endless pursuit of more stimulation. It is not simply a way to occupy our time, it's a way to cultivate deeper joy, deeper satisfaction, and deeper appreciation of the present moment.

Neuroscience of Pornography Addiction

And, as we move through this process of re-engagement with the world we also begin to recognize the power of choice, and the ability to choose to pause before acting. In the cycle of addiction, individuals often feel as though they are compelled by their urges, as if they have lost control of their actions, and are simply being pulled along by the force of their cravings. However, as we begin the process of recovery, we begin to realize that there is always a space between stimulus and response, a space where we have the power to choose. The ability to pause before acting is the key to breaking free of old habits, and this space allows us to become more aware of our thoughts, our feelings, and our impulses before acting upon them. This pause is not simply about waiting, it's about recognizing that we do have the power to decide, to choose, and to act with intention. This moment of choice is often overlooked but it can, over time, become the most powerful tool in breaking the addictive cycle.

Recognize that by consciously choosing our actions we begin to reclaim agency and choice. The experience of

Kendir Ramiz

addiction often feels like a loss of control, as though we are being driven by forces outside of our own conscious awareness. However, as we progress through recovery, we begin to understand that we are not merely passive victims of our urges, but that we are capable of making conscious choices, and of intentionally directing our own lives. Reclaiming agency, is about recognizing our own capacity to choose, and to make decisions that align with our values, our goals, and our overall sense of well-being. This is not simply about stopping the addiction, it's about consciously choosing the type of life we want to live, and the types of experiences that we want to create. It's about taking the reins of our own life, and deciding that we are the masters of our own destiny. We can choose to not be a passive participant in the addiction cycle, but to become the architect of our own experiences. The very act of choosing empowers us to change our responses, to change our patterns and to change the trajectory of our lives.

Reframing pleasure and desire is a critical step in the recovery journey. It is about moving beyond instant

Neuroscience of Pornography Addiction

gratification to find sources of pleasure that are sustainable and fulfilling, appreciating the value of effort and achievement, engaging in activities mindfully, and recognizing the power of choice in breaking free from old habits. The journey from addiction to recovery is not simply about denying ourselves pleasure; it's about intentionally cultivating a more meaningful relationship with pleasure, one that is rooted in awareness, intentionality, and a deeper appreciation for the richness of human experience. The very capacity to choose, to pause, and to act with intention is a powerful force that can completely transform our lives, and it is that force that can pull us from the depths of addiction into the freedom of an intentional and more meaningful way of being. This is not merely a reframing of pleasure; it is a reclaiming of life itself.

Kendir Ramiz

Beyond the Science

Having explored the intricate details of the neurochemical processes involved in addiction, the various pathways of recovery, and the importance of reframing pleasure and desire, we now arrive at the final chapter, a space for reflection, integration, and the acknowledgment that the journey of healing extends far beyond the realm of science alone. This chapter is dedicated to exploring the ongoing nature of recovery, recognizing that it is not a destination to be reached but a continuous process of growth, change, and self-discovery. It's as if we are acknowledging that life is not a static event, but a dynamic and ever-evolving process, and our understanding of ourselves must also be continually evolving. This is not about reaching a state of perfection, but about embracing the imperfections, and recognizing the power of transformation that is present in every moment. It's about recognizing that the scientific aspects of recovery, as important as they are, can only take us so far. What we must also acknowledge, is that we, as human beings, are so much more than a collection of chemical processes, and our healing journey must reflect that deep interconnectedness of our physical, mental, emotional and spiritual lives. This is not simply the end of the book; it is the beginning of a new chapter, a chapter that is still unwritten, and full of potential.

Let us begin with the crucial understanding that recovery is not a destination but an ongoing process. This is perhaps one of the most important things to understand about any journey of healing, and that is, that it does not end at some magical moment, but rather, it continues to evolve throughout a lifetime. It is not about reaching a state of permanent sobriety, or a state where we never struggle again; it is about learning to navigate the complexities of life with greater awareness, greater resilience, and a greater capacity for self-compassion. It is about understanding that the challenges, and the obstacles, that come our way, are not a sign that we have failed, but simply another opportunity to learn, grow, and evolve. Recovery, in this sense, is not a linear process, it's not that we move from one phase to another without ever revisiting the previous one; it is often a circuitous path, with ups and downs, with moments of clarity and moments of confusion, and with moments of joy and moments of sorrow. This also means that relapses are not a sign of failure but, as we have already explored, they are often a natural part of the process, and they can be used as opportunities for greater self-understanding and growth. The journey of recovery, in other words, is a lifelong journey, a process of continuous adaptation, learning, and refinement, not something that ends at a specific point. It is a dance, a give and take, and it is in that dance that we often find deeper wisdom, and a deeper sense of self.

And, central to that continuous journey, is the importance of cultivating self-compassion and self-forgiveness. It's as if we recognize that we are not perfect, and that the journey of recovery will be filled

Neuroscience of Pornography Addiction

with challenges, setbacks, and times where we struggle, and in those times we must meet ourselves with understanding. Self-compassion, is about treating ourselves with the same kindness, understanding, and empathy that we would offer to a friend, and to recognize that we are all doing the best we can with what we have been given. Self-forgiveness is about releasing the self-blame, shame, and guilt that often accompany addiction, and to embrace our humanity, in all of its imperfections. This means acknowledging our past actions without dwelling on them, and to forgive ourselves for the choices that we have made, understanding that we were also doing the best we could at that particular time. This is not about condoning our mistakes, it's about recognizing that we are all human, and as such, we are prone to errors and missteps, and that those very mistakes are often opportunities for growth and self-discovery. It is not simply about letting go of guilt, but about embracing ourselves with all of our flaws and our imperfections. The journey of recovery is not a path towards perfection, it's a journey toward greater self-acceptance.

Kendir Ramiz

Resilience, is not simply about being able to bounce back after a setback, it's about our capacity to adapt, to learn from our experiences, and to keep moving forward, even when things get difficult, and those challenges are not an indication of our weakness, but often a clear sign of our ability to adapt and grow. Hope, is the belief that a better future is possible, that our efforts will be rewarded, and that we have the strength and the capacity to overcome the challenges that we face. Hope and resilience are not simply abstract concepts; they are powerful forces that fuel our commitment, inspire our efforts, and provide us with the strength to keep going when we feel like giving up. In the context of recovery, resilience and hope are not simply about a positive attitude, they are about a deeply ingrained belief in ourselves, in our capacity for transformation, and our ability to create a life that is free from the cycle of addiction. These qualities are not something we are born with, they are something we must develop over time, they must be cultivated, and nurtured. When we begin to embody resilience and

Neuroscience of Pornography Addiction

hope, we recognize that there is always another path forward, that even in the midst of challenges, we are always capable of change.

And as we begin to find those other paths, let's explore the transformative potential of turning pain into purpose. The experiences of addiction can be incredibly challenging, often filled with suffering, pain, loss, and despair. However, those very experiences can also serve as a powerful source of wisdom, empathy, and motivation. This is about taking the difficult moments and allowing them to serve as a foundation for creating a life that is filled with meaning, and purpose. When we understand that our difficulties can be a catalyst for growth, we begin to recognize that we are not merely victims of our circumstances, we can also become the architects of our own destiny. This is not simply about finding a way to cope with the pain; it's about using the very lessons learned through that pain, to help others, to contribute to something larger than ourselves, and to find a deeper sense of meaning in the process. It's as if we are recognizing that even the darkest moments, can contain the seeds of transformation, and growth. This is the essence of turning pain into purpose, using one's lived experience as a source of inspiration for others.

The enduring power of the human spirit, a force that transcends all of the scientific explanations, and exists as a force in and of itself. This is about recognizing our innate capacity for healing, growth, and transformation, and to recognize that this capacity exists within us, and that we are often more powerful than we even realize. The human spirit, is not just an abstract concept; it's a vital force, a wellspring of resilience, a source of hope, and a deep and abiding connection to life itself. It is the inner spark that drives us to seek, to grow, and to evolve. It is the force that can propel us from the depths of addiction, and into the light of healing, wholeness, and genuine well-being. In the context of recovery, the human spirit is not simply something that we observe, it is something we embody, something we nurture, and something we come to rely upon. This is not just a nice idea, it's a fundamental truth: our capacity for transformation is far greater than we can even imagine.

Neuroscience of Pornography Addiction

This is a reminder that we are not just biological beings; we are also spiritual beings, and our capacity to overcome even the greatest challenges, is inherent within our very nature.

The journey of healing extends far beyond the realm of science, it is an ongoing process of growth, and change; it requires self-compassion and self-forgiveness; it is fueled by resilience and hope, it can transform pain into purpose and it is rooted in the enduring power of the human spirit. The path to recovery is not a simple one, but by understanding the complex interplay of the body, the mind, and the spirit we can create a life that is filled with greater meaning, greater joy, and greater fulfillment. The journey of healing is not simply about overcoming an addiction; it's about embracing our full potential as human beings, and in so doing we may find a deeper sense of peace, purpose, and a deep and unwavering belief in the power of human transformation. This is not merely the end of a book, it is the beginning of a journey, a journey that is yours, and yours alone, and one that is filled with endless possibilities.

Kendir Ramiz

Reiterating the Importance of Scientific Understanding

As we draw near the end of this extensive exploration into the complex world of pornography addiction, it becomes paramount to reiterate a fundamental truth that has underscored our journey from the very beginning: the indispensable power of knowledge. This is not merely about collecting facts and figures, or about memorizing scientific terminology; it is about embracing a deeper understanding of the intricate mechanisms that shape our behavior, our desires, and our struggles. Scientific knowledge, in this context, is not an abstract academic pursuit; it is a practical, transformative tool that empowers us to break free from the chains of ignorance and to actively participate in our own healing and growth. It's as though we are acknowledging that the first step in any significant transformation is always understanding, and that a deeper, more nuanced understanding of ourselves and the forces that drive our behaviors, is the very first step on the path towards liberation. This isn't simply about believing in science; it's about recognizing the ways that scientific inquiry provides practical and valuable tools that can help us make better choices and live more fully.

The journey we have undertaken in this book has been grounded in the principles of scientific inquiry, a process that is rooted in observation, analysis, and a commitment to evidence-based reasoning. We have explored the neurochemical processes that underlie the initial spark of desire, the formation of habits, the dysregulation of the reward system, and the body's complex attempts to counter those imbalances. We have delved into the roles of various neurotransmitters, hormones, and brain structures, recognizing that these seemingly abstract processes have a very real and direct impact on our lived experiences. We have seen how the intense stimuli of pornography can hijack the brain's reward system, creating patterns of compulsive behavior that are incredibly difficult to break. And through that exploration, we have come to understand that addiction is not simply a matter of willpower or moral failing; it is a complex neurological phenomenon, that is deeply intertwined with our own biological makeup and conditioned behavioral responses. This is not to say that we are simply products of our biology, but rather, that our biology is often a powerful influence on our behaviors, and that by understanding those influences we can begin to move beyond the limitations of those biological processes.

This scientific understanding is not just valuable for researchers and clinicians, it is essential for anyone seeking to navigate the complexities of addiction, either for themselves or for others. When we understand the underlying mechanisms of addiction, we move beyond the simplistic and often unproductive approaches that are rooted in shame, guilt and moral judgement. We begin to see that the struggle is not simply about a lack of willpower, or a character flaw, but a very real neurological process, and those neurological processes

Neuroscience of Pornography Addiction

can be addressed. By understanding how the brain responds to addictive stimuli, we can develop more effective strategies for breaking free from the cycle of compulsion, and to build a more conscious relationship with our desires. This scientific literacy also empowers us to advocate for better treatment options, to challenge stigma, and to create a more compassionate and understanding environment for those who are struggling, and that environment may be the very thing that someone needs in order to finally be able to heal. This knowledge is not just for experts; it is knowledge that is valuable for all of humanity, and for all those who wish to understand the human condition more fully.

This scientific approach allows us to move beyond the simplistic notion that addiction is a choice. As we have explored at length in this book, the brain, when exposed to addictive stimuli, undergoes very real changes, and when those systems are hijacked the individual is often left with very little ability to make choices that serve them well. These changes impact reward circuitry, impulse control, and even the very ability to feel pleasure in response to natural stimuli. Understanding these processes allows us to recognize that addiction is not a matter of lacking self-discipline or willpower; it's a complex condition that often requires comprehensive treatment approaches that target both the neurological and psychological aspects of the problem, and it is through that recognition that we move into a space where real healing can begin to occur. The shame, the guilt, and the moral condemnation that often accompany addiction simply serve to reinforce the cycle and prevent the person from seeking out help. A scientific understanding of addiction challenges this mindset and creates space for compassion, empathy, and a greater understanding for those who struggle. It is in this very recognition that we begin to shift towards a more understanding and compassionate paradigm.

This scientific understanding also provides a framework for developing more effective solutions. When we know how the brain responds to addictive stimuli, we can design interventions that target specific neural pathways, that help to rewire the brain, and that support the processes of recovery. These interventions may include pharmacological treatments, that assist with managing withdrawal symptoms, that reduce cravings and enhance the body's own healing processes; therapeutic approaches, like cognitive behavioral therapy (CBT), mindfulness-based stress reduction (MBSR), or other forms of therapy that help individuals to address underlying traumas, build coping mechanisms, and reframe their relationship with desire; or lifestyle interventions, that include good nutrition, adequate sleep, and regular exercise, which help to restore balance and support brain health. The key is to approach addiction with a comprehensive and multi-faceted plan, that acknowledges both the neurological, psychological and social factors that contribute to the problem. This is not simply about treating a set of symptoms; it's about addressing the root cause of the problem and creating a foundation for lasting health and wellbeing.

Scientific literacy has another vitally important role, it helps us to challenge the stigma and shame that often surround addiction. When we understand addiction as a complex neurobiological phenomenon, we can move beyond judgement and blame, and create a more understanding and compassionate environment for those who are struggling. This includes advocating for better treatment options, promoting open and honest conversations about addiction, and creating a culture where people feel safe to seek help without fear of judgement or condemnation. This is not just about being kind and compassionate; it's also about creating conditions that support actual change. When individuals feel ashamed and stigmatized they often hide their struggles, which often prevents them from accessing the resources they need in order to get better. When we understand addiction as a health issue, not a moral failing, we can then work together to create a society that empowers individuals to seek out help, to heal and to thrive. It is in this understanding that we move from shame and secrecy to empathy and acceptance.

The power of knowledge, and more specifically scientific knowledge, helps us to recognize our capacity for

Neuroscience of Pornography Addiction

change. The brain is not a static structure; it's a dynamic and ever-changing system that is capable of remarkable adaptation. The very concept of neuroplasticity, the brain's capacity to rewire itself, offers a tangible, scientific basis for the hope that is inherent in all paths of recovery. The ability to understand how the brain learns, adapts, and changes empowers us to take an active role in creating our own healing journey. This scientific literacy is not simply about understanding the problem; it's about recognizing our own capacity for change and intentionally and consciously creating a path towards our goals. It's about recognizing that we are not simply victims of our biology; we are also the architects of our own lives.

The power of knowledge, especially scientific knowledge, is an indispensable tool in navigating the complexities of pornography addiction and formulating effective solutions. Scientific knowledge provides a framework for understanding the underlying causes of addiction, creating effective treatment plans, challenging stigmas and empowering individuals on their path to healing. This is about understanding that knowledge is not simply a collection of abstract facts and figures, it is a powerful force that can help us change our lives, our communities, and our relationship with the human condition. We have journeyed far together, and in that journey we have seen that by applying this scientific understanding we begin to move towards a greater understanding of addiction, and a greater capacity to heal. This is not simply about knowing, it is about transforming that knowledge into practical actions, and ultimately creating a more compassionate and understanding world. The path towards healing begins with knowledge, and in that knowledge we often find hope, inspiration, and the strength to continue on our journey.

Illuminating the Path Forward

As we bring this comprehensive exploration of the neuroscience of pornography addiction to a close, it becomes imperative to emphasize that our understanding of this complex phenomenon is far from complete. We have made significant strides in recent years, but there remain vast areas yet to be explored, questions yet to be answered, and mysteries yet to be unraveled. This is not a call to despair; it is a call to action, a plea for continued scientific inquiry, and a demand for open, honest conversations that can illuminate the path forward, both for those who are struggling with addiction and for society as a whole. This isn't about declaring victory and closing the book; it's about recognizing the ongoing nature of scientific discovery and the importance of continued inquiry as we seek to understand more fully the nuances of the human experience. It is as though we are acknowledging that

Kendir Ramiz

our journey of learning is never truly finished, that we must continue to ask questions, seek out knowledge, and deepen our understanding of ourselves, and the forces that shape our behaviors. This is not just about a thirst for knowledge; it is about a deep and abiding commitment to truth, and to the ongoing exploration of all of the complexities of life.

The field of neuroscience is in constant evolution, with new discoveries being made on a regular basis, and our understanding of the brain, its intricate circuitry, and its dynamic interactions with the environment is ever-evolving. While we have gained significant insights into the neurochemical processes involved in addiction, there remain many questions that demand further investigation. What are the precise mechanisms by which different individuals respond to pornography? How do genetic predispositions, early childhood experiences, and cultural factors influence the development of addictive behaviors? What are the most effective strategies for preventing addiction, for intervening early, and for promoting long term recovery?

Neuroscience of Pornography Addiction

How can we better understand and address the co-occurring mental health conditions that often accompany addiction? These are not simply academic questions, they are real issues, and the answers to these questions have profound implications for individuals, families, and society as a whole. This isn't merely an academic inquiry; it's a practical need, and we must recognize that there is still so much that we do not know, and that ongoing research is essential to deepen our understanding, and to develop more effective solutions.

One of the crucial areas that requires further exploration is the impact of pornography on the developing brain. Adolescence and young adulthood are critical periods of neural development, and it is essential that we understand the ways in which exposure to pornography may alter these developmental trajectories. How does chronic pornography use affect the prefrontal cortex, the area of the brain responsible for higher-order cognitive functions? Does early exposure to pornography impact the development of healthy sexual attitudes and behaviors? How does the brain's reward system develop differently when it is exposed to hyper-stimuli from an early age? These questions are not simply about the long term impacts, they are also about the ethical considerations that must be at the forefront of all of these conversations. We have a responsibility to protect the developing minds, and we must begin to understand and address the potential impact that this constant exposure to technology, and pornography is having on the younger generation. This is not simply about curiosity, it's about responsibility. It is a call to action, and a plea for further research.

We need to delve deeper into the ways in which individual differences influence the development and course of addiction. Why do some individuals become addicted to pornography, while others do not? How do genetic predispositions, environmental factors, and psychological vulnerabilities interact to create susceptibility to addiction? What are the most effective strategies for personalized treatment, those that can be tailored to the specific needs of the individual? These questions highlight the complex interplay between nature and nurture, between our genetic inheritance, our life experiences, and the environment that shapes us. Understanding these complexities will allow us to move beyond one-size-fits-all approaches, and to develop more effective and individualized treatments that address the root cause of addiction. This isn't simply about finding a better treatment, it's about recognizing the profound differences between individuals, and understanding that our approaches to treatment must be as complex and nuanced as the human condition itself.

And beyond those biological factors, we must also begin to explore the social, cultural and ethical dimensions of pornography addiction. How does societal messaging around sexuality, the objectification of the human body, and the pervasive nature of online access contribute to the problem? How do different cultural norms shape the experience and perception of pornography? What are the ethical implications of easily accessible, often exploitative, and increasingly realistic digital content? These questions underscore the fact that pornography

Kendir Ramiz

addiction is not simply a biological issue; it is also a cultural, social, and ethical one. Addressing this complex issue requires a broader understanding of the context in which addiction develops, and of the ways in which society reinforces and perpetuates harmful behaviors. This isn't simply about the individual's choices, it is about the collective responsibility that we all have, to create a healthier, more equitable, and more compassionate society. It's about understanding that our behaviors, our values and our actions all contribute to the overall health of society, and in so doing, we must take responsibility for creating the world that we wish to see.

Alongside the call for more research, we must also emphasize the need for open and honest conversations about pornography addiction. This is not a topic that should be relegated to the shadows of shame and secrecy; it is an issue that affects individuals, families, and communities across the globe. Open conversations, that are rooted in scientific understanding and compassionate understanding can help to destigmatize addiction, reduce shame, and empower individuals to seek out help without fear of judgement or condemnation. These conversations can also help us to develop more effective prevention strategies, and that begin in the home, and continue in the classroom, with the ultimate goal of creating a culture where healthy relationships and informed decision-making are highly valued. This is not simply about starting a debate; it's about fostering a dialogue that is rooted in respect, empathy, and the shared commitment to create a better future. It is through these open and honest conversations that we begin to understand that we are not alone, and in that understanding we may find the strength, and the courage to keep moving forward.

These conversations must also include input from those who have direct lived experience of pornography addiction. Often those who have walked the path of addiction and recovery, provide unique insights that can enrich our scientific understanding and guide the development of more effective treatment and prevention strategies. Their wisdom, and their lived experiences are often invaluable and we, as a society, must create space for their voices to be heard. This is not just about including their perspectives, it is about empowering them to become active participants in the research and conversation. Their lived experience provides a unique type of knowledge that is not always captured in scientific data, and we must not overlook the wisdom that comes from the path of direct experience.

The call for more research and awareness is a call to collective action. It is a recognition that we must work together—scientists, clinicians, educators, policy makers, and members of the community—to address the challenges of pornography addiction and to create a society that supports health, well-being, and human flourishing. It's about taking responsibility for the world we are creating, and for ensuring that we are creating an environment where our children and future generations have every opportunity to live a life that is free from the limitations of addiction. This is not simply a problem for individuals to solve on their own, it is a collective responsibility. And in that shared responsibility we also have an opportunity to collectively create a better world.

The call for ongoing research and awareness is not simply a plea for more information; it's a commitment to a future where we understand the nature of desire and addiction more fully, and where we have the tools to create a more compassionate, equitable, and healthier society. This is a call to challenge our assumptions, to question our beliefs, and to engage in a continuous process of learning, and growth. The journey toward greater understanding is never truly complete, and that journey is one that we must continue to walk together. It is about continuing the conversation, and in so doing, continuing to create a future that is more just, more equitable, and more supportive of the human spirit. This is not merely a call for more knowledge, it is a call for greater humanity.

Kendir Ramiz

Fostering a Healthier Relationship with Technology and Each Other

As we approach the culmination of this extensive exploration into the neuroscience of pornography addiction, it is critical to underscore that the challenge before us is not merely an individual one; it is a collective responsibility that demands a profound shift in our societal values, our cultural norms, and our relationship with technology. This is not just about understanding the science of addiction; it's about recognizing that we are all interconnected, that our actions, our choices, and our collective consciousness shape the world in which we live, and therefore, we must collectively acknowledge our role in fostering both the problem, and the solution. It's as though we are recognizing that we are all part of a larger ecosystem, and that the health of the entire system is dependent

Neuroscience of Pornography Addiction

upon the health of each individual component. This is not simply a matter of personal choices, it's about understanding that our choices have an impact far beyond our own personal lives, and by embracing this responsibility, we can move towards a more conscious, more compassionate, and ultimately more sustainable way of living. This is not just a call to action; it is a call to collective awareness, a recognition that we must all participate in the healing of our communities, and our societies.

We, as a global society, have become inextricably intertwined with technology. The digital age has transformed nearly every facet of our lives, from how we communicate, to how we work, to how we entertain ourselves, and how we learn, and it is in that constant state of connection that we must begin to see the impact on our collective well-being. Technology, like fire, is a powerful tool; it can be used to create, to connect, and to empower, but it can also be used to destroy, to manipulate, and to isolate. We have witnessed the profound benefits of technology, but we must also acknowledge the potential harms and the very real challenges that arise when technology becomes a dominating force in our lives. The ubiquitous nature of technology, combined with the increasing sophistication of persuasive design, has created a landscape where we are all, in a sense, being constantly influenced by external forces that are beyond our immediate control. And this requires that we act consciously and intentionally to address the potential risks, and to cultivate a more healthy relationship with our technological devices. This is not about demonizing technology, but about recognizing its inherent power, and understanding the responsibility that comes with that power.

The ubiquity of pornography in the digital age is a clear example of the ways in which technology can be used to exploit our vulnerabilities and to create conditions that can contribute to addictive behaviors. The ease of access, the anonymity, and the hyper-stimulating nature of pornography has created a perfect storm that has overwhelmed many of the natural regulatory

Neuroscience of Pornography Addiction

mechanisms of the human brain, creating cycles of dependence, isolation, and a loss of self-control. And we, as a society, must take responsibility for the impact that this has had on both individuals, families, and communities. This is not simply about policing online content, or controlling access to these types of materials; it's about addressing the underlying societal issues that have contributed to this problem, and creating a culture where healthy relationships, authentic connections, and ethical consumption of media is highly valued. It is in this deeper understanding that we may begin to address this problem in a way that is sustainable, ethical, and effective. This is not simply a problem for the individuals to solve on their own, it is also an indication that we must all take responsibility for the world we have collectively created, and actively participate in the healing of our communities.

We must also acknowledge our shared responsibility in creating a culture that values authentic connection and healthy relationships. Addiction, as we have explored throughout this book, often thrives in isolation, and is a direct result of a lack of genuine connection, and when we create communities that are rooted in empathy, compassion and mutual support, we make it far more difficult for addiction to take root. This requires a conscious effort to foster open and honest communication about difficult topics, to create spaces for vulnerability and authenticity, and to prioritize human interaction over technological stimulation. It's as if we are recognizing that we are not just individuals seeking our own forms of pleasure, we are also social creatures that long for connection, belonging and to be seen, heard and known for who we truly are, and when we create spaces for these needs to be met, we reduce the isolation that is so often the foundation for addictive behaviors. We must understand that we are not simply separate individuals, but rather, we are all part of a larger web of relationships, and when that web is strong, we all become stronger.

This collective responsibility also extends to the ways in which we educate our children, and how we equip them with the tools to navigate the complexities of the digital

Neuroscience of Pornography Addiction

age. We must begin to foster a healthy relationship with technology at a young age, teaching children how to use these tools responsibly, ethically, and consciously. We must empower them to develop critical thinking skills, emotional regulation, and the capacity to discern between healthy and unhealthy forms of engagement. We must also provide open and honest conversations about sexuality, relationships, and the potential harms of pornography, and in so doing, create a culture where they are empowered to make conscious, and informed choices. It is our responsibility to prepare the younger generation for the challenges of the future, and to ensure that they have the tools they need to live healthy and fulfilling lives. We must recognize that our children are the future of our world, and we have a responsibility to protect them, nurture them, and empower them to create a society that is more just, more compassionate, and more supportive of human flourishing.

This also includes a call to action for policymakers and for leaders across all of our communities, to recognize the complex nature of addiction, and to develop policies that are evidence-based, ethical, and compassionate. This includes creating more accessible treatment options, promoting prevention programs in schools and communities, and challenging the harmful stigmas and stereotypes that often surround addiction. This also includes taking responsibility for ensuring that technology is used ethically, that the corporations that are building these tools, are held accountable for their actions, and that the long term well-being of individuals and society, is always a top priority. This is not simply about placing limits on technology, it's also about using technology to support our goals, our values, and to create a more just, and more equitable world for all. It's as if we are acknowledging that we, as a society, have the power to shape the future, and therefore we must all take responsibility for creating a world that supports health, happiness and human dignity.

It is important that we acknowledge that this is not just about addressing the problem of pornography addiction; it's about creating a society that is healthier, more just, and more compassionate for all. This requires

Neuroscience of Pornography Addiction

addressing a wide range of social issues, from poverty, to inequality, to access to mental health care. We must also acknowledge the ways that our economic and social systems often reinforce patterns of exploitation, isolation, and despair, and we must work to create a more equitable and supportive environment for all members of society. It is also important to acknowledge that systemic change is difficult, it requires us to challenge the status quo, to question our assumptions and to work together towards creating a better world. This is not simply about solving addiction; it's about recognizing the fundamental interconnections that shape our lives.

This is a call to recognize that we are all in this together, that our individual well-being is inextricably linked to the well-being of the collective, and that when we take responsibility for ourselves, for our communities and for our shared future, that we may begin to create a better world for all. This is not simply about addressing the problem of pornography addiction; it's about creating a society that supports the full potential of the human spirit, and it is in that realization that we begin to heal not just ourselves, but our communities, and our world. This is not merely a call to action; it's a call to consciousness, to compassion, and to collective responsibility. It's a call to not only address the issues, but to create a society that is rooted in empathy, respect, and a genuine commitment to the well-being of all. This is about recognizing that the problem is us, and therefore, the solution must be us as well. It's not about waiting for others to act; it's about each of us taking personal responsibility for creating the world that we want to live in.

Fostering a healthier relationship with technology and with each other is not simply an individual pursuit; it's a collective responsibility that demands a profound shift in our societal values, our cultural norms, and our relationships with each other. By taking responsibility for the impact of technology, by prioritizing human connection, by educating our children, and by supporting ethical policy-making, we can create a world that is more compassionate, more just, and more conducive to human flourishing. This is not simply about addressing the specific problem of pornography addiction; it's about creating a society that supports the well-being of all, and recognizes the interconnected nature of human existence. This is not just a problem for some of us to solve; this is a challenge for all of us to face together. And in that togetherness we may find the strength, the wisdom, and the compassion to create a world that is more reflective of our highest ideals.

Kendir Ramiz

The Potential for Transformation

As we arrive at the final moments of this extensive exploration into the complexities of pornography addiction, it is essential to reinforce a message of enduring hope: the profound and ever-present potential for transformation. This is not simply a platitude or a comforting sentiment; it is a fundamental truth grounded in both scientific understanding and the enduring capacity of the human spirit. Recovery from addiction is not a predetermined outcome, nor is it a miraculous event; it is an ongoing journey of growth, change, and self-discovery, a path that is available to all who are willing to embark upon it. It is as though we are acknowledging that our capacity to learn, adapt, and evolve, is an inherent part of the human experience, that even in our darkest moments, the possibility for

Neuroscience of Pornography Addiction

transformation remains, waiting to be awakened. This is not just about overcoming addiction; it's about discovering the hidden depths of our own potential, and to step into the fullness of who we are meant to be. This isn't simply about escaping the limitations of our past; it's about intentionally creating a future that is rooted in freedom, integrity, and authentic self-expression.

We have delved into the intricate details of the neurochemical processes involved in addiction, and we have explored the ways in which the brain responds to hyper-stimulating stimuli. We have seen the development of habits, the dysregulation of reward systems, and the challenges of withdrawal, and through all of that we have always maintained an awareness that this is not the complete picture, that there is something beyond these biological processes, and that human beings are so much more than the sum of their parts. We have seen how addiction can create cycles of dependence, isolation, and despair, but we have also explored the pathways of recovery, highlighting the power of neuroplasticity, the importance of self-compassion, and the role of human connection. And through all of that, there has been an undercurrent of hope, and a deep-seated belief that transformation is not just possible, but that it is an inherent part of the human experience. This is not simply a blind faith; it is a recognition of the very real capacity that we all have to change, to grow, and to evolve.

The concept of neuroplasticity, the brain's ability to reorganize itself by forming new neural connections, offers a tangible scientific basis for this enduring hope. The brain is not a fixed entity; it is a dynamic and ever-changing system that is constantly adapting to new experiences. The very fact that the brain can rewire itself in response to learning, is a profound reminder that even patterns of behavior that feel deeply entrenched, can be changed with intentionality, effort, and the dedication to move beyond the confines of the past. This means that the negative changes in brain structure and function that result from chronic addiction are not permanent; they can be reversed through conscious effort and healthy choices. This is not to say that it is

Neuroscience of Pornography Addiction

always easy or effortless, but it does mean that we are not bound to the limitations of our past experiences, and that we have the capacity to create new neural pathways that support healthy behaviors, and fulfilling lives. The brain's inherent ability to heal is a testament to the enduring power of the human spirit, and to the endless possibilities for change and growth.

We must also recognize that recovery is not simply about the cessation of addictive behavior; it is often a deep process of transformation that involves all aspects of an individual's life. This process often includes addressing underlying emotional issues, healing past traumas, rebuilding damaged relationships, and cultivating new values, and new perspectives on life. This is not just about stopping a particular behavior, it is also about developing a sense of purpose, meaning, and connection with something larger than oneself. The journey of recovery can often lead to a greater sense of self-awareness, self-compassion, and a more authentic expression of oneself, and is in that very transformation, that we discover the power of the human spirit. It's as if

Kendir Ramiz

we are acknowledging that we are not simply seeking an escape from something, but we are also actively creating a future that is rooted in our deepest values and our highest potential. This is not just about eliminating the old patterns; it's about intentionally choosing the new patterns that are more sustainable, more satisfying, and more fulfilling.

And within this transformation, we often find that the very struggles, challenges, and difficulties of the past become powerful sources of wisdom and compassion. It is as if we are recognizing that even the most challenging moments are often opportunities to grow, to learn, and to gain a greater understanding of ourselves and the world around us. The pain and hardship associated with addiction, while incredibly difficult, can be a catalyst for profound personal growth, and those experiences can empower us to support others, to offer empathy, and to become advocates for change in our communities. The ability to transform pain into purpose, is a powerful force that is inherent in the human spirit, and in that transformation we often discover that we are

Neuroscience of Pornography Addiction

capable of doing far more than we ever imagined possible. This is not about glorifying the struggles, it's about acknowledging their transformative potential, and in so doing, to find the wisdom that is often hidden within the dark and challenging moments of life. It's as if we are recognizing that even the most challenging experiences can be used to create a more compassionate and understanding world.

And this transformation is not a solitary journey; it is one that is often enriched by the support, the compassion, and the wisdom of others. Human connection is a vital part of this process, and the ability to share our stories, to be seen and heard by others, and to connect with a community of people who are also on the path to recovery often allows us to move forward with greater strength, greater courage, and a deeper sense of belonging. The journey of transformation is not one that we must take alone; it is one that we can embrace together, and in that togetherness we often find the support we need to move towards our goals, to embrace our full potential, and to create a more meaningful and fulfilling life. We are, after all, social beings, and in those connections we often find the wisdom that we require to move forward on our path of healing.

The potential for transformation is a reflection of the enduring power of the human spirit. It's as though we are recognizing that our innate ability to adapt, to grow, to heal, and to overcome, is stronger than any addiction, any limitation, and any obstacle that we face. The human spirit is not a fragile thing; it is resilient, it is adaptable, and it is capable of remarkable feats of strength, and courage, and it is in that realization that we begin to move with a more enduring sense of hope, and a greater capacity to trust in our own power to create lasting change. This understanding allows us to move from a state of powerlessness to a state of empowerment, from a state of hopelessness to a state of enduring belief, and a state of limitation into a state of endless possibilities. It is not simply about changing our behavior, it is about transforming our very sense of who we are, and the world that we inhabit. This is not a journey of defeat; it is a journey of discovery, a journey

Neuroscience of Pornography Addiction

of profound transformation, and a journey that is open to all who are willing to embrace it.

Recovery from pornography addiction is not merely a cessation of compulsive behavior; it is a profound journey of transformation, growth, and self-discovery. The brain's inherent capacity for neuroplasticity offers a tangible scientific basis for this hope, and that this capacity to change is one that we must continually nurture and believe in. The journey of recovery is a process of reclaiming one's life, of rebuilding relationships, and cultivating a deeper sense of self-worth and purpose. The challenges and difficulties of the past can become powerful sources of wisdom and compassion, empowering individuals to connect with others and to become advocates for change. The potential for transformation is inherent in the human spirit, and it is always available to us when we choose to embrace it, when we choose to believe in ourselves, and when we choose to take responsibility for our own paths. It is not simply about finding a solution, it is about transforming the problem into something that has the potential to help us grow, evolve, and live with greater purpose and greater meaning. And it is in this understanding that we may all find the enduring hope that is necessary to move forward, and to embrace the endless possibilities that are available to us all. This isn't merely a conclusion; it's a call to embrace our potential, and step into the fullness of who we are meant to be.

Looking Forward with Hope and Action

As we reach the culmination of this extensive exploration into the intricate landscape of pornography addiction, it is essential to conclude not with a sense of finality, but with a resounding call to action. This is not a moment to rest on our laurels, or to simply bask in the knowledge that we have gained; it is a time to embrace the potential for a brighter future, a future where we continue to deepen our understanding, where we translate that understanding into effective solutions, and where we collectively commit to creating a world that supports the well-being of all. This is not just about addressing the challenges of addiction; it's about envisioning a new horizon, where we move beyond the limitations of the past, where we embrace the complexities of the present, and where we work together to build a more compassionate, just, and fulfilling future.

It's as if we are standing at a crossroads, acknowledging that our journey thus far has led us to this very moment, and it is from this moment, that we must look towards a future that is not just possible, but also one that is worthy of our highest aspirations. This is not simply a conclusion; it's an invitation to join a movement, and to actively participate in creating a world that reflects our deepest values, and our most enduring hopes.

Throughout this book, we have journeyed through the intricate details of the neuroscience of pornography addiction, and we have explored the underlying mechanisms that drive compulsive behaviors, and the ways that the brain responds to the hyper-stimulating world in which we live. We have examined the complex interplay of neurotransmitters and hormones, the power of reward pathways, and the devastating impact of excitotoxicity, and all of this has illuminated the complexity of this challenge that we are collectively facing. We have delved into the importance of addressing the physical, mental, emotional and social dimensions of recovery, and we have also highlighted

Neuroscience of Pornography Addiction

the power of human connection, self-compassion, and the capacity for transformation. And in all of this, we have also recognized that our understanding is ever-evolving, and that the process of discovery is one that we must continually embrace. This is not just about recognizing what we know, it is also about acknowledging what we don't know, and committing ourselves to a continuous path of exploration and understanding.

Now, as we move forward, it is imperative that we translate this understanding into concrete actions. This requires a multi-faceted approach that targets the underlying causes of addiction, that challenges the systemic issues that perpetuate the problem, and that empowers individuals to reclaim their lives and to build more fulfilling futures. This is not just a call for more scientific research; it's a call for the application of knowledge to create a real, measurable and enduring impact in the world. This means developing more effective treatment and prevention strategies, providing accessible and affordable mental health services, and creating educational programs that promote healthy relationships, and responsible technology use. It also means addressing the social and economic inequalities that contribute to vulnerability to addiction, creating an environment that promotes opportunity, equity, and compassion for all. This isn't just about addressing addiction; it's about building a better world for everyone. This calls for a comprehensive approach, that sees that we are all part of the same interconnected web of life, and in so doing, what impacts one, has an impact on all.

It is also important that we recognize that the path forward requires a collective commitment. This is not a challenge that can be addressed solely by individuals or by isolated groups; it requires a unified effort that includes researchers, clinicians, policymakers, educators, community leaders, and all members of society. We must create spaces for open dialogue, collaborative action, and shared responsibility, and when we do so, we begin to build a movement of change, transformation and healing, that has the power to create a more sustainable, equitable and just future. This is not

Neuroscience of Pornography Addiction

just about agreeing on a set of solutions; it's about recognizing that we are all partners in creating a world that supports the well-being of all. This also requires that we each take ownership of the role we play, both in creating the problem, and in creating a solution. This is not about pointing fingers, it's about extending hands, and creating the collective force that is necessary to create profound change.

And, in that shared collaboration, we must also remain steadfast in our commitment to evidence-based practices. This means relying on scientific research, data-driven approaches, and the insights that are gained from those with lived experience, and to be willing to challenge our own assumptions, and to constantly be seeking to learn more. In a world filled with so much misinformation, and so many conflicting points of view, it's critical that we maintain our connection with truth, and to continue to refine our understanding of addiction and recovery through the rigorous methods of scientific inquiry. This is not simply about adhering to the scientific method; it's about cultivating a mindset of intellectual humility, a willingness to adapt and evolve in light of new discoveries, and an ongoing desire to seek a greater understanding of the world around us, and the complex ways in which that world is influencing our behavior. This means moving beyond dogmatic beliefs, and staying flexible and open to change, always with an eye towards the truth, and the best possible outcomes.

It is also equally important that we remain grounded in empathy and compassion. While the scientific approach allows us to understand the underlying mechanisms of addiction, it is critical that we also remember that behind every addiction there is a person who is struggling, a person who is experiencing pain, and a person who deserves our compassion. This is not just about understanding their challenges; it's about connecting with their humanity, their strength and their enduring potential. This means offering a non-judgmental space for people to share their stories, to feel heard, and to receive the support that they need. It also means challenging the stigma and shame that often surround addiction, and creating a culture where seeking help is

Neuroscience of Pornography Addiction

seen as a sign of strength, and not as an indication of weakness. This is not simply about changing our behaviors; it's about changing the world, and our relationship to each other, creating spaces where people feel safe, seen, heard and loved for who they truly are.

We must recognize that the path forward will not always be easy, that there will be obstacles, challenges, and setbacks, and we will at times feel like we are losing ground. The complexity of addiction and the scale of the problem is significant, and it requires that we remain steadfast in our commitment, even in the face of adversity. The journey of transformation is never linear, and there will always be unexpected challenges along the way, and in those times we must remember that resilience, persistence and the power of the human spirit are vital forces for creating positive and enduring change. It's as if we are acknowledging that the challenges we face, are often the very opportunities that allow us to grow, to adapt, and to become stronger, both as individuals and as a collective. This is not about denying the difficulties; it's about embracing those challenges as an opportunity to grow.

We also must remain grounded in the hope that a brighter future is not only possible but that it is also inevitable, when we commit ourselves to action, and when we move forward with an enduring belief in the power of the human spirit, and in the infinite capacity that we all have to grow, learn, and evolve. This is not simply an optimistic sentiment; it is a recognition that we have within ourselves the power to create a world that is rooted in well-being, that is just and compassionate, and where all individuals have the opportunity to live fulfilling lives that are rooted in purpose, meaning and authentic connection. The future is not simply something that happens to us; it is something that we create through our choices, our actions and our beliefs. We have the potential to create a future that is far better than the past, and by working together, by embracing science and by choosing compassion we can move into a world that has more possibilities than we can even imagine. This is not just a belief; it is a deep knowing, and it is

Neuroscience of Pornography Addiction

from this knowing that we must now move forward with courage, with hope, and with a deep commitment to action.

The path forward requires that we combine both a deep scientific understanding with a deep commitment to action, that we embrace our shared responsibility, and that we never lose sight of the enduring power of hope. By continuing to learn, by applying that knowledge, and by creating a movement for change, we can create a future where the challenges of pornography addiction are met with wisdom, compassion, and effective solutions. The power to create a brighter future lies within us all; it is time for us to embrace that potential, and to embark upon the journey with courage, with conviction, and with a deep and abiding belief in the transformative power of human potential. This is not simply a call for a better future; it is a call for each of us to become the architects of that future, and to move forward with a deep and abiding belief in the power of collective transformation. The future is not something that is destined to happen, it is something that we

Kendir Ramiz

create, and we now have the opportunity to shape that future and in so doing to create a better world for all. This is not a closing statement; it's a launching point, and it is time for us to embark upon the next chapter of our collective journey.

www.ingramcontent.com/pod-product-compliance
Lightning Source LLC
Chambersburg PA
CBHW052211220526
45471CB00004B/1908